教育部人文社会科学研究青年基金项目（20YJCZH144）
广东省基础与应用基础研究基金面上项目（2019A1515010884） 资助
广东省珠江人才计划高层次人才项目（20170133）

唐 凯 著

气候变化与雨养农业
——基于微观证据与国际比较的生物经济学分析

Climate Change and Rainfed Agriculture:
a Bioeconomic Analysis Based on
Micro-evidence and International Comparison

中国财经出版传媒集团

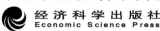

图书在版编目（CIP）数据

气候变化与雨养农业：基于微观证据与国际比较的生物经济学分析／唐凯著．—北京：经济科学出版社，2021.6

ISBN 978-7-5218-2601-2

Ⅰ.①气… Ⅱ.①唐… Ⅲ.①气候变化－影响－农业－研究 Ⅳ.①S343.1

中国版本图书馆 CIP 数据核字（2021）第 109320 号

责任编辑：杜　鹏　常家凤
责任校对：杨　海
责任印制：王世伟

气候变化与雨养农业
——基于微观证据与国际比较的生物经济学分析

唐　凯　著

经济科学出版社出版、发行　新华书店经销
社址：北京市海淀区阜成路甲 28 号　邮编：100142
编辑部电话：010-88191441　发行部电话：010-88191522
网址：www.esp.com.cn
电子邮箱：esp_bj@163.com
天猫网店：经济科学出版社旗舰店
网址：http://jjkxcbs.tmall.com
固安华明印业有限公司印装
710×1000　16 开　13.75 印张　240000 字
2021 年 6 月第 1 版　2021 年 6 月第 1 次印刷
ISBN 978-7-5218-2601-2　定价：69.00 元
（图书出现印装问题，本社负责调换．电话：010-88191510）
（版权所有　侵权必究　打击盗版　举报热线：010-88191661
QQ：2242791300　营销中心电话：010-88191537
电子邮箱：dbts@esp.com.cn）

前　言

越来越多的科学证据表明，地球正在经历着显著的气候变化，而且这种变化在日益加速。纵观历史，地球一直在经历由自然原因所致的气候变化，而现在我们所经历的变化与以往不同，人类活动是造成目前气候变化的主要原因。

气候变化是当今国际社会普遍关注的一个全球性议题。近年来，随着社会经济的不断发展，气候变化给人类社会所带来的负面影响日趋显现，深刻影响着人类生存和发展。为了应对气候变化给全球带来的重大挑战，国际社会开始采取联合行动来抑制温室气体的排放，共同应对气候变化。这是全球可持续发展的内在需要，也是推动构建人类命运共同体的责任担当。

习近平总书记多次强调，应对气候变化不是别人要我们做，而是我们自己要做，是中国可持续发展的内在要求，是主动承担应对气候变化国际责任、推动构建人类命运共同体的责任担当。应对气候变化是推动中国经济高质量发展和生态文明建设的重要抓手，是参与全球治理和坚持多边主义的重要领域，事关中国发展的全局和长远。在第七十五届联合国大会一般性辩论上，习近平指出，应对气候变化《巴黎协定》代表了全球绿色低碳转型的大方向，是保护地球家园需要采取的最低限度行动，各国必须迈出决定性步伐。同时，就中国新达峰目标与碳中和愿景向国际社会作出庄严承诺。这彰显了中国积极应对气候变化、走绿色低碳发展道路的雄心和决心，为各国携手应对全球性挑战、共同保护好人类赖以生存的地球家园贡献了中国智慧和中国方案，受到了国际社会的广泛认同和高度赞誉。习近平总书记宣布的新达峰目标与碳中和愿景，是党中央、国务院统筹国际国内两个大局作出的重大战略决策，对当前和今后一个时期乃至21世纪中叶的应对气候变化工作、绿色低碳发展和生态文明建设提出了更高要求，有利于促进经济结构、能源结构、产业结构转型升级，有利于推进生态文明建设和生态环境保护、持续改善生态环境质量，对于加快形成以国内大循环为主体、国内国际双循环相互促进的新发展格局,推动高质量发展,建设美丽中国,具有重要促进作用。

雨养农业是世界上最大的农业生产部门,使用了全球 3/4 以上的耕地,是全球超过 1/2 人口的食物来源。同时,雨养农业也是受气候变化影响最深远最直接的生产部门。气候变化增加了全球雨养农业区农业生产的不确定性,直接影响了农业生产布局和结构。因此,有必要分析气候变化所带来的自然环境以及社会经济等方面的变化对雨养农业的影响,并在此基础上探寻雨养农业对气候变化的适应性路径。这是合理预测未来可持续发展必不可少的工作,对于包括中国在内的世界各国特别是广大发展中国家应对气候变化、保障农业生产与粮食安全、减少农村贫困、促进区域环境经济协同发展有着重要意义,也是当前学界较少涉及的一个新领域。本书正是基于这样的背景而开展的。

本书从雨养农业的具体特点和实际出发,基于已有的科学研究成果和实地观测数据,系统梳理雨养农业区气候变化的中长期趋势及其潜在变化幅度,分析气候变化对雨养农业生产的影响;通过国际比较,从微观的视角出发,分析雨养农业温室气体减排的成本有效性,探索在优化农户收益前提下雨养农业对气候变化的有效适应路径,为应对气候变化、保障农业生产与粮食安全、推动农业绿色发展与转型、促进区域环境经济协同发展提供了理论依据和政策参考。书中部分研究结果已先期发表在一些国际学术期刊。相关图表如无说明,均为作者制作或拍摄。

本书关于气候变化对雨养农业的影响以及雨养农业对气候变化的有效适应方面的分析,进一步补充和完善了生物经济学、农业经济学、环境经济学、自然地理学以及系统科学在农业可持续发展理论上的认识,对丰富和延伸气候变化适应研究的内容和深度、明晰雨养农业生产系统各部分的内在联系、规范农业可持续性适应研究范式有着重要的理论和实践意义。

人类共享同一个地球。我们在某一个地方做出改变,会影响相距遥远的其他人。因此,在采取行动和做出选择的过程中,每个人都可以通过自身的努力,为应对气候变化留下点点足迹。

唐 凯

2021 年 5 月

目 录

第1章　绪论 ·· 1
　1.1　研究背景及意义 ·· 1
　1.2　研究目标 ··· 23
　1.3　研究方法 ··· 24
　1.4　创新点 ··· 27

第2章　文献综述 ··· 29
　2.1　气候变化条件下雨养农业的相关生物物理学基础 ············· 29
　2.2　气候变化所引起的制度环境方面的改变 ·························· 39
　2.3　气候变化对于雨养农业的影响 ··· 46
　2.4　雨养农业对于气候变化的适应 ··· 51
　2.5　气候变化对雨养农业的影响及其适用的相关研究方法 ······ 56
　2.6　国内外相关文献的总结性评述 ··· 59

第3章　理论基础 ··· 61
　3.1　农业系统理论 ··· 61
　3.2　外部性理论 ·· 63
　3.3　公共物品与产权理论 ··· 66

第4章　澳大利亚雨养农业温室气体的减排潜力和成本 ········ 70
　4.1　引言 ··· 70
　4.2　距离函数及其求解方法述评 ·· 74
　4.3　研究方法与数据 ·· 82
　4.4　研究结果与讨论 ·· 88
　4.5　本章小结 ··· 95

第 5 章 澳大利亚雨养农业种—畜复合经营家庭农场对温室气体减排政策的响应 …… 97

- 5.1 引言 …… 97
- 5.2 相关研究述评 …… 99
- 5.3 研究对象与方法 …… 101
- 5.4 研究结果 …… 109
- 5.5 进一步讨论 …… 118
- 5.6 本章小结 …… 122

第 6 章 黄土高原地区雨养农业温室气体减排的成本有效性分析 …… 124

- 6.1 引言 …… 124
- 6.2 相关研究述评 …… 128
- 6.3 研究对象与研究方法 …… 130
- 6.4 研究结果 …… 136
- 6.5 进一步讨论 …… 141
- 6.6 本章小结 …… 144

第 7 章 黄土高原地区雨养农业小农户对气候变化影响的适应 …… 145

- 7.1 引言 …… 145
- 7.2 相关研究述评 …… 147
- 7.3 研究对象与分析方法 …… 149
- 7.4 研究结果 …… 153
- 7.5 进一步讨论 …… 158
- 7.6 本章小结 …… 161

第 8 章 结论、建议与展望 …… 163

- 8.1 全书总结 …… 163
- 8.2 政策建议 …… 165
- 8.3 研究展望 …… 167

附录 …… 170

参考文献 …… 179

第1章 绪　　论

1.1 研究背景及意义

1.1.1 研究背景

1.1.1.1 气候变化：人类社会共同面临的挑战

气候变化（climate change）是指全球或区域气候的平均状态在一段时间内的变动，包含气候持续较长一段时间（通常是 30 年或以上）的变动或者在统计意义上的显著偏离其长期平均状态的现象。气候变化是由地球内部运动（如造山运动）、外部力量（如太阳辐射变化），以及人类活动（如温室气体排放）共同作用所引起的。[①] 以全球变暖和极端气候频发为主要特征的气候变化是当今国际社会普遍关注的全球性议题。近年来，随着社会经济的不断发展，气候变化给人类社会所带来的负面影响日趋显现。

全球气候变暖是气候变化的一个主要特征。已有研究显示，19 世纪中期以来，全球大气表层温度上升了 0.6℃，是近 1000 年以来最剧烈的变化。2012 年与 1880 年相比，全球表面平均温度上升了 0.85℃（见图 1-1）。2006~2015 年与 1850~1900 年相比，全球表面平均温度升高了 0.75℃~0.99℃，陆地表层大气温度升高了 1.38℃~1.68℃（见图 1-2）。受强厄尔

[①] IPCC. Managing the Risks of Extreme Events and Disasters to Advance Climate Change Adaptation. A Special Report of Working Groups I and II of the Intergovernmental Panel on Climate Change [R]. Cambridge University Press, Cambridge, UK, 2012.

尼诺事件影响的2016年是有完整气象记录以来最暖的一年，比工业化前基线高1.2℃，比20世纪中期平均值高0.99℃。① 2019年，全球平均温度较工业化前水平高出约1.1℃，是有完整气象观测记录以来的第二暖年份。2015～2019年是有完整气象观测记录以来最暖的5个年份（见图1-3）；20世纪80年代以来，每个连续10年都比前一个10年更暖。2019年，亚洲陆地表面平均气温比常年值（使用1981～2010年气候基准期）偏高0.87℃，是20世纪初以来的第二高值。预计到21世纪中期，全球表面平均温度将上升1.5℃。全球许多地区出现了更为明显的区域气候变暖，其中，占全球人口20%～40%的一些地区经历了至少一个升温超过1.5℃的季节（IPCC，2018）。

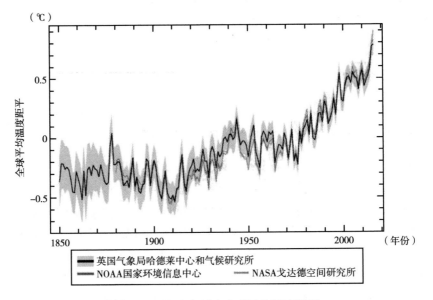

图1-1　1850年以来全球平均温度距平

资料来源：世界气象组织（https：//public.wmo.int/zh-hans/media/新闻通稿/2016年的气候打破多项纪录，影响波及全球）。

近百年来，中国年平均地面气温升高1.15℃（见图1-4），升高速率为0.10℃/10年，与全球大陆平均增温趋势接近。20世纪中期以来，中国年平均气温上升趋势越发明显，1951～2019年中国年平均气温上升速率达到0.24℃/10年，升温速率明显高于同期全球平均水平。2018年中国平均气温

---

① https：//www.nasa.gov/press-release/nasa-noaa-data-show-2016-warmest-year-on-record-globally.

较常年偏高 0.54℃。① 近 20 年是 20 世纪初以来的最暖时期。

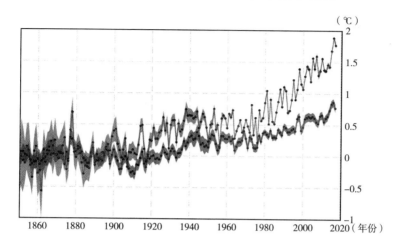

图 1-2　1850~2012 年所观测到的全球陆地和海洋表面平均温度异常

资料来源：陆地温度数据来自 Berkeley Earth，海洋温度数据来自 HadSST。

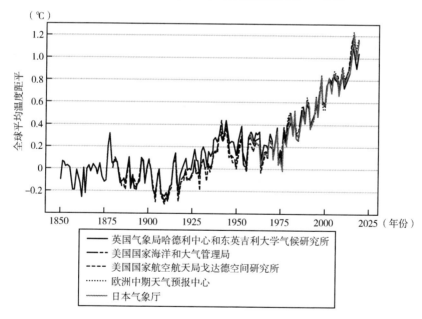

图 1-3　1850~2019 年全球平均温度距平（相对于 1850~1900 年平均值）

资料来源：中国气象局气候变化中心：《中国气候变化蓝皮书（2020）》，2020。

---

① 中国气象局气候变化中心：《中国气候变化蓝皮书（2020）》，2020。

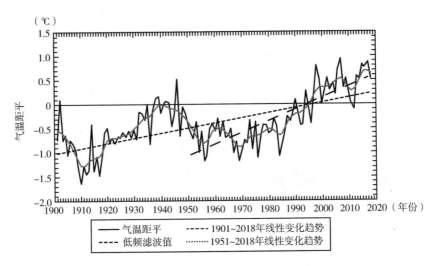

图 1-4　1901~2018 年中国地表年平均温度距平

资料来源：中国气象局气候变化中心：《中国气候变化蓝皮书（2020）》，2020。

全球气候变暖很有可能造成了冰川、冰帽和极地冰盖的融化，从而导致了全球海平面的逐渐上升。[①] 1901~2010 年，全球平均海平面上升了 190 毫米。20 世纪以来，全球平均海平面上升了超过 200 毫米。自 1993 年以来，全球平均海平面上升的速度均值为 3.1 毫米/年（见图 1-5）。[②] 自 1979 年起，北极的海冰范围以 107 万平方千米/10 年的速度持续缩小。鉴于当前温室气体的浓度以及排放水平，21 世纪末全球平均气温将持续升高，高出前工业化时期的平均水平；世界各大洋将持续变暖，冰雪将继续融化。以 1986~2005 年作为参照期，至 2065 年，平均海平面预计上升 240~300 毫米；至 2100 年，平均海平面预计上升 400~630 毫米。即使停止排放温室气体，气候变化所带来的大多数影响也会持续数世纪之久。

中国沿海海平面变化总体呈波动上升趋势。1980~2019 年，中国沿海海平面上升速率为 3.4 毫米/年，高于同时段全球平均水平。过去 10 年中国沿海平均海平面处于近 40 年来高位（见图 1-6）。2020 年，中国沿海海平面较

[①] IPCC，2007：《气候变化 2007：综合报告．政府间气候变化专门委员会第四次评估报告第一、第二和第三工作组的报告》，[核心撰写组、Pachauri，R. K 和 Reisinger，A.（编辑）]，IPCC，瑞士日内瓦，第 2 页。

[②] Jackson, J., Choudrie, S., Thistlethwaite, G., et al. UK Greenhouse Gas Inventory, 1990 to 2007: Annual Report for submission under the Framework Convention on Climate Change [R]. UK Department of Energy and Climate Change, 2009.

1993～2011年平均值高72毫米，为1980年以来第三高（见图1-7），较2018年升高25毫米。渤海、黄海、东海和南海沿海海平面与常年相比分别高86毫米、60毫米、79毫米和68毫米。①

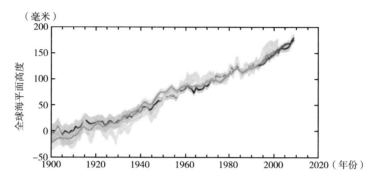

图1-5　1900～2012年与全球海平面高度变化

资料来源：政府间气候变化专门委员会（IPCC）（https：//www.ipcc.ch/site/assets/uploads/2018/02/SYR_AR5_FINAL_full_zh.pdf）。

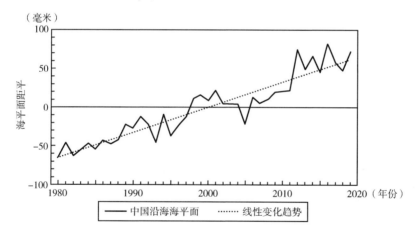

图1-6　1980～2019年中国沿海海平面距平（相对于1993～2011年平均值）
资料来源：中国气象局气候变化中心：《中国气候变化蓝皮书》，2020。

与此同时，全球极端气候频发。② 高温、干旱、降水时空分布异常等现

---

① 自然资源部海洋预警司：《2019年中国海平面公报》，2020。
② 极端天气事件（extreme weather event）是指在某一特定区域短时间内发生的未能预期到的、严重的、罕见的天气事件。极端天气事件包含极端高温（如热浪）、极端低温（如寒潮）、强降水（如大暴雨、暴雪）和强热带气旋（如飓风、台风、热带风暴）等。如果一种类型的极端天气持续了较长一段时间，如一个季节或连续几年都发生，则可被归为极端气候事件（extreme climate event）。

象出现的频次和强度显著上升（见图1-8）。联合国政府间气候变化专门委员会（Intergovernmental Panel on Climate Change, IPCC）在其公布的《应对

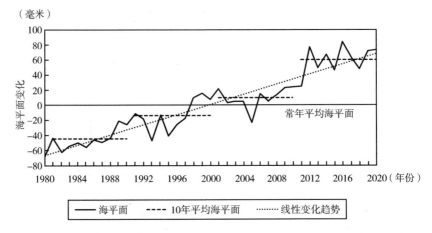

图1-7　1980~2020年中国沿海海平面变化（相对于1993~2011年平均值）

资料来源：自然资源部海洋预警司：《2020年中国海平面公报》，2021。

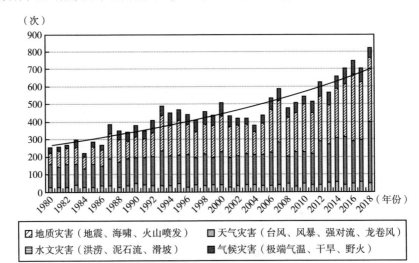

图1-8　1980~2018年全球重大自然灾害发生次数

注：自然灾害事件入选的标志为至少造成1人死亡或至少造成10万美元（低收入经济体）、30万美元（下中等收入经济体）、100万美元（上中等收入经济体）或300百万美元（高收入经济体）的损失；经济体划分参考世界银行相关标准。

资料来源：谢伏瞻、刘雅鸣：《应对气候变化报告（2019）：防范气候风险》，社会科学文献出版社2019年版。

极端事件与灾害风险以推进对气候变化的适应》报告①中指出，自20世纪50年代以来，北美洲、欧洲以及澳大利亚的大部分地区出现高温天气的频率以及高温天气持续的天数有非常明显的增加。自20世纪中叶以来，在南欧以及西非地区，干旱严重程度以及持续时间呈增加态势。在地中海地区、中欧地区、北美洲中部地区、中美洲和墨西哥、巴西北部以及南部非洲地区，干旱愈加频发。在全球大部分地区（特别是北美洲），强降水天气出现的频率呈现上升态势。在季风区，极端降水天气更加多发，区域性洪水发生的概率有所增加。

中国气象局在《中国气候变化蓝皮书（2019）》和《中国气候变化蓝皮书（2020）》中指出，中国极端高温事件自20世纪90年代中期以来明显偏多（见图1-9），2019年云南元江（43.1℃）等64站日最高气温达到或突破历史极值。极端日降水量事件的频次明显增加，1961~2019年，中国平均年降水量呈微弱的增加趋势，平均年降水日数呈显著减少趋势，极端强降水事件呈增多趋势，年累计暴雨（日降水量≥50毫米）站日数呈增加趋势，平均每10年增加3.8%。登陆中国台风比例自2000年以来增加趋势明显，登陆台风平均强度自20世纪90年代后期以来呈现明显的偏强特征。自20世纪90年代初期以来，中国气候风险指数明显增高，1991~2018年平均气候风险指数较1961~1990年平均值增加了54%。②

气候变化引起的极端天气每年都给中国造成了巨大的损失。2014年以来，中国极端天气频发，南方多地出现强降雨，城市内涝、局部洪涝、山洪、滑坡、泥石流等灾害大幅增加。超强台风"威马逊"是1949年以来登陆中国的最强风暴，③给海南、广东、广西等地造成了严重的人员及财产损失。北方多地出现极端高温事件。云南、河南等地出现重度干旱，极大地影响了当地农业生产和人民生活。自2020年5月下旬起，在长江中下游地区、淮河流域、西南、华南及东南沿海等地因持续强降水引发了严重洪灾。截至7月28日，主汛期（6月1日）以来，洪涝灾害造成江西、安徽、湖北等27省（区、市）5481.1万人次受灾，158人死亡失踪，376万人次紧急转移安置；

---

① IPCC. Managing the Risks of Extreme Events and Disasters to Advance Climate Change Adaptation. A Special Report of Working Groups I and II of the Intergovernmental Panel on Climate Change [R]. Cambridge University Press, Cambridge, UK, 2012.

② 中国气象局气候变化中心：《中国气候变化蓝皮书》，2019。

③ https://www.chinanews.com/gn/2014/07-25/6427488.shtml.

4.1万间房屋倒塌,36.8万间不同程度损坏;农作物受灾面积528.33万公顷;直接经济损失1444.3亿元。①

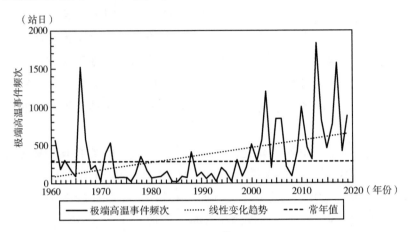

图1-9 中国极端高温事件频次

资料来源:中国气象局气候变化中心:《中国气候变化蓝皮书(2020)》,2020。

气候变化对全球自然生态和人类系统构成了严峻的挑战。联合国政府间气候变化专门委员会在其公布的《气候变化2007:综合报告》中指出,自1980年以来,高温所引起的疾病或死亡的总体风险稳步攀升,而全球约有30%的人口目前生活在持续极端热浪的气候条件下。2000~2016年,全球易受热浪事件影响的脆弱人口数量增加了约1.25亿。报告指出,2016年,2350万人在天气相关灾害期间被迫迁移。与往年一样,这些境内迁移绝大多数都与洪水或风暴相关,并发生在亚太地区。在东亚地区,预计到2050年,大的江河流域其可用的淡水会减少,众多三角洲地带会面临严重的洪水威胁,自然资源和环境的压力会严重加剧。在澳大利亚,一些生态资源丰富地区的生态多样性受到显著损失;到2030年,其东部和南部水安全问题会加剧,农业和林业产量会因为干旱和火灾的增多而下降;到2050年,由于海平面上升、飓风以及洪水的严重影响,海岸地带的发展所面临的风险将增大;到2100年,全球平均地表气温升高约3.5℃的情况下,全球总产量将降低7%~14%,在热带和贫困地区的损失甚至更大。因气温升高引起的生产率损失而带来的社会成本将分别在2020年、2030年增加至73~142美元/tCO_2e、

---

① 中华人民共和国应急管理部,https://www.mem.gov.cn/xw/bndt/202007/t20200728_354105.html。

92~181美元/tCO$_2$e。①

有证据表明,生态系统和地球气候系统可能已经达到甚至突破了重要的临界点,可能导致不可逆转的变化,这让人担忧。亚马孙雨林和北极苔原等多样化的生态系统可能因气候变暖和干旱而发生巨大的变化。高山冰川正在迅速消失,在最干旱的月份里,供水减少对下游造成的影响会波及很多世代。如果当前国际社会不采取紧急行动,未来要适应这些影响会变得更加困难,成本也会更加高昂。

目前学界对于气候变化的具体机理和诱因尚存争议。但是绝大多数学者都认为,由人类活动排放的温室气体(greenhouse gases)很可能是全球气候变暖的重要原因。温室气体是自然产生的,可以阻挡部分太阳光反射回太空,使得地球温度适合生物居住,对人类以及其他数以百万计物种的生存至关重要。然而,在经历了150多年的工业化、砍伐森林和大规模的农业生产之后,大气中温室气体的含量增长到了300万年来前所未有的水平。随着人口的增长、经济的发展和生活水平的提高,温室气体排放总量也随之增加。

自工业化时代以来,人类活动排放的温室气体快速增加。② 大气中含量最多的温室气体、约占其总量2/3的二氧化碳(CO_2),主要由人类焚烧化石燃料产生。过去100年间大气中温室气体的浓度达到了以往2.2万年来的最高峰,1970~2004年排放量增加了70%,其中,二氧化碳的排放增加了约80%(见图1-10)。③ 2005年大气中二氧化碳和甲烷的浓度远超过了过去65万年的自然变化的范围。④ 全球大气中二氧化碳的浓度在2011年为391ppm,相较于工业化前时代增加了约40%;2017年进一步上升到405ppm,比1750年前后水平上升了46%。2018年全球二氧化碳平均浓度水平为407.8ppm,相当于工业化前水平(1750年之前)的147%,与2017年的水平相比进一步增长,达到过去80万年的最高水平。连续30年(1985~1995年、1995~

---

① Kalkuhl, M., Wenz, L. The impact of climate conditions on economic production. Evidence from a global panel of regions [J]. Journal of Environmental Economics and Management, 2020, 103, 102360.
② IPCC.《气候变化2007:综合报告.政府间气候变化专门委员会第四次评估报告第一、第二和第三工作组的报告》,日内瓦:IPCC,2007年。
③ IPCC.《气候变化2007:综合报告.政府间气候变化专门委员会第四次评估报告第一、第二和第三工作组的报告》,日内瓦:IPCC,2007年,第5页。
④ 王小彬、武雪萍、赵全胜等:《中国农业土地利用管理对土壤固碳减排潜力的影响》,载《中国农业科学》2011年第11期。

2005年和2005~2015年），二氧化碳的平均增长率从1.42ppm/年增至1.86ppm/年再增加到2.06 ppm/年，厄尔尼诺事件期间观测到最高的年增长率。

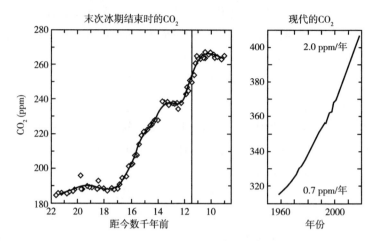

图1-10　全球大气中二氧化碳浓度变化

资料来源：世界气象组。https://public.wmo.int/en/media/press-release/greenhouse-gas-concentrations-surge-new-record.

甲烷（CH_4）是第二重要的长寿命温室气体，贡献约17%的辐射强迫。大约40%的甲烷通过自然源（如湿地和白蚁）排入大气，约60%的排放来自人类活动，如畜牧、水稻种植、化石燃料利用、垃圾填埋和生物质燃烧等。甲烷的浓度在2016年为1853ppb，相较于工业化前时代增加了约157%。2018年大气甲烷创下新高，约为1869ppb，目前是工业化前水平的259%。2017~2018年的甲烷增加值为过去10年来的历史第二高（见图1-11）。

氧化亚氮（N_2O）是通过自然源（约60%）和人为源（约40%）排入大气，包括海洋、土壤、生物质燃烧、化肥的使用以及各种工业流程。大气中氧化亚氮的浓度在2016年为328.9ppm，相较于工业化前时代增加了约22%。[①] 2018年大气中氧化亚氮的浓度为331.1ppb，是工业化前水平的123%。2017~2018年的增长量高于2016~2017年的观测值以及过去10年的平均增长率（见图1-12)[②]。

[①] https：//public.wmo.int/en/media/press-release/greenhouse-gas-concentrations-surge-new-record.

[②] https：//public.wmo.int/zh-hans/media/新闻通稿/大气温室气体浓度再创新高.

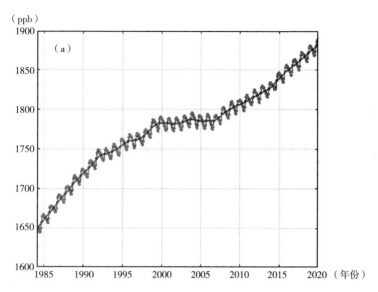

图 1–11 1984~2019 年全球平均甲烷摩尔分数

注：曲线为去除季节变化后的月平均值；点和折线表示月平均值。

资料来源：引自世界气象组（WMO）。https：//library.wmo.int/doc_num.php？explnum_id=10463.

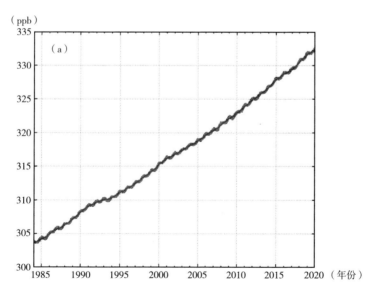

图 1–12 1984~2019 年全球平均氧化亚氮摩尔分数

注：曲线为去除季节变化后的月平均值；点和折线表示月平均值。

资料来源：引自世界气象组（WMO）。https：//library.wmo.int/doc_num.php？explnum_id=10463.

全球大气温室气体平均浓度在2017年为501.5ppm碳当量（CO$_2$e）。① 2019年和2020年，主要温室气体的浓度持续上升。根据联合国环境规划署的数据，经济衰退暂时抑制了新的温室气体排放，但对大气中温室气体的浓度没有明显的影响。工业化时代温室气体、气溶胶排放以及土地利用与土地覆盖的变化极大地改变了大气成分，进而影响了地球能量平衡，引起了当今的气候变化②。如果现有的温室气体排放速度保持不变，那么到21世纪末全球平均气温将上升约5℃甚至更多。③

为了应对气候变化给全球带来的挑战，国际社会开始采取联合行动来抑制温室气体的排放。1992年，联合国召开了第一次有世界各国首脑参加的国际气候变化会议。作为会议的成果，与会各方同意签署《联合国气候变化框架公约》（United Nations Framework Convention on Climate Change，UNFCCC）。该公约的目的在于将大气中温室气体的浓度控制在一个能够避免危险的气候变化的稳定水平上。④ 在1992年共有166个国家和地区签署了这项公约，截至2017年10月，缔约方增加到189个。

作为联合国气候变化框架公约的补充条款，签订于1997年的《京都议定书》（The Kyoto Protocol）规定，世界上主要的37个工业化国家以及欧盟成员国需要在2008～2012年控制其温室气体排放量至少比1990年的水平减少5%。⑤ 绝大多数发展中国家虽然签署了京都议定书，但不被强制要求减少温室气体排放量。纵观工业革命以来的历史，发达国家应当为大气中温室气体浓度的增加负主要责任。然而，随着一些发展中大国经济社会的快速增长，未来全球温室气体的排放将主要来自发展中国家。有预测显示，发展中国家

---

① https://www.csiro.au/en/Research/OandA/Areas/Assessing-our-climate/State-of-the-Climate-2018/Greenhouse-gases.

② IPCC. Managing the Risks of Extreme Events and Disasters to Advance Climate Change Adaptation. A Special Report of Working Groups I and II of the Intergovernmental Panel on Climate Change［R］. Cambridge University Press，Cambridge，UK，2012. 塔西甫拉提·特依拜、丁建丽：《土地利用/土地覆盖变化研究进展综述》，载《新疆大学学报（自然科学版）》2006年第1期。

③ Government of India. Assessment of climate change over the Indian Region. A report of the Ministry of Earth Sciences（MoES），Government of India［R］. Springer Nature Singapore，Singapore，2020.

④ United Nations Framework Convention on Climate Change（UNFCCC），Kyoto Protocol（United Nations；Germany），A decision support framework for greenhouse accounting on Australian dairy，beef or grain farms［R］. UNFCCC，2009.

⑤ Flannery，T.，Beale，R.，Hueston，G.，The critical Decade：International Action on Climate Change［R］. Commonwealth of Australia Department of Climate Change and Energy Efficiency，2012.

温室气体排放量占全球总量的比重将从 2001 年的 44% 上升到 2025 年的 53%。① 截至 2012 年（京都议定书原定截止时间②），除了加拿大，其他《京都议定书》签署国已基本完成其做出的减排承诺。③

2009 年的哥本哈根世界气候变化大会做出决议，要求发达国家承诺减少温室气体的排放量，而发展中国家要采取措施逐步抑制温室气体排放量的增长速度，以确保未来全球平均温度的上升幅度不高于 2℃。④ 有研究显示，要想达到这一目标，到 2050 年，全球温室气体的排放量至少要比 1990 年的水平下降 50%。⑤

在 2010 年的坎昆世界气候变化大会上，包括发达国家和发展中国家在内的 90 个国家承诺采取措施减少或者抑制其温室气体排放量，并重申了确保未来全球平均温度的上升幅度不高于 2℃ 的重要性。这些国家每年排放温室气体的总量占全球排放总量的 80% 以上。⑥

2011 年的德班世界气候变化大会达成协议，要求缔约方在 2015 年之前达成一项具有法律约束力新协议，并在 2020 年正式实施。美国和中国作为世界上最大的温室气体排放国，同意参与旨在减少全球温室气体排放的全球协议。

2012 年底召开的多哈世界气候变化大会就京都议定书第二阶段、发达国家与发展中国家依照联合国气候变化框架公约进行长期合作行动等重要问题达成了一系列的重要协议，并推动了为落实德班气候变化大会决议而进行的谈判。

在 2015 年底召开的巴黎世界气候变化大会上，联合国气候变化框架公约近 200 个缔约方达成了一项历史性的协议，即《巴黎协定》。该协定主要包括以下内容。

（1）把全球平均气温升幅控制在工业革命前水平不超过 2℃ 的范围内，并努力将气温升幅限制在工业革命前水平不超过 1.5℃ 范围内，这将大大降

---

① https://www.bloomberg.com.au/apps/news?pid=20601087&sid=aduzrUpnfTeM&pos=8.
② 2012 年在卡塔尔召开的联合国气候变化大会上，京都议定书的到期时间被延长到 2020 年。
③ The Climate Institute, Policy Brief-Australia and the Future of the Kyoto Protocol [R]. The Climate Institute, 2012.
④⑤ DAFF. Carbon farming initiative brochure [R]. Australian Department of Agricultural, Fisheries and Forestry, 2012.
⑥ Smith, P., Martino, D., Cai, Z., et al. Greenhouse gas mitigation in agriculture [J]. Philosophical Transactions Royal Society of London Biological Sciences, 2008, 363: 789-813.

低气候变化所造成的不利影响和风险。

（2）提高适应气候变化所带来的不利影响的能力，并以不损害粮食生产的方式增强气候抗御力和温室气体低排放发展。

（3）使资金流动符合温室气体低排放和气候适应性发展的路径。

2016年的马拉喀什世界气候变化大会就《巴黎协定》程序性议题达成共识，各缔约方宣布进入履行《巴黎协定》行动期。与会各方欢迎发达国家就2020年前实现每年提供1000亿美元资金支持提出路线图，并呼吁发达国家继续增加可用资金，以最终兑现承诺。

2017年的波恩气候变化大会通过了名为"斐济实施动力"的一系列成果，就《巴黎协定》实施涉及的各方面问题形成了谈判案文，进一步明确了2018年促进性对话的组织方式，通过了加速2020年前气候行动的一系列安排。

2018年的卡托维兹气候变化大会取得了一揽子全面、平衡、有力度的成果。参会各方就《巴黎协定》关于自主贡献、减缓、适应、资金、技术、能力建设、透明度、全球盘点等涉及的机制、规则基本达成共识，并对下一步落实《巴黎协定》、加强全球应对气候变化的行动力度提出进一步安排。大会成果体现了公平、共同但有区别的责任、各自能力原则。大会成果传递了坚持多边主义、落实《巴黎协定》、加强应对气候变化行动的积极信号，提振了国际社会合作应对气候变化的信心。

2019年的马德里气候变化大会就一些重要的问题达成一致，例如能力建设、性别计划和技术。但由于在一些更重大、更具争议性的问题上未能达成一致，全面的协议未能出台，这些事宜有关解决人为造成的气候变化导致的损失和破坏，以及气候适应方面的融资问题。在会议期间，欧盟宣布致力于到2050年实现碳中和，73个国家和地区宣布他们将提交增强的气候行动计划；14个地区、398座城市、786家企业和16个投资方正在努力到2050年实现二氧化碳净零排放；177家公司同意确保公司业务符合科学家的倡议，将全球升温幅度限制在1.5°C以内，并在不迟于2050年达到净零排放的水平。①

应对气候变化是人类共同的事业。中国政府始终高度重视应对气候变化。近年来，习近平总书记多次强调，应对气候变化不是别人要我们做，而是我

---

① https://news.un.org/zh/story/2019/12/1047431.

们自己要做,是中国可持续发展的内在需要,也是推动构建人类命运共同体的责任担当。习近平总书记在全国生态环境保护大会上明确提出,要实施积极应对气候变化国家战略,推动和引导建立公平合理、合作共赢的全球气候治理体系。

考虑到温室气体减排对于人类社会和我国应对气候变化以及推进可持续发展的重要性,在2009年哥本哈根世界气候大会上,中国政府向世界宣布:中国到2020年单位国内生产总值温室气体排放量比2005年下降40%~45%,并将其作为国民经济与社会发展中长期计划中的一项约束性指标。2014年11月中国国家主席习近平与美国时任总统奥巴马在北京共同发布的《中美气候变化联合声明》中宣布,中国计划2030年左右二氧化碳排放达到峰值且将努力早日达峰,并计划继续努力并随时间而提高力度。国家发展和改革委员会于2017年6月公布的《"十三五"控制温室气体排放工作方案部门分工》中明确指出,"十三五"期间累计减排二氧化碳当量11亿吨以上,减少农田氧化亚氮排放,到2020年实现农田氧化亚氮排放达到峰值。2020年9月22日,习近平主席在第七十五届联合国大会一般性辩论上发表重要讲话强调,中国将提高国家自主贡献力度,采取更加有力的政策和措施,二氧化碳排放力争2030年前达到峰值,努力争取2060年前实现碳中和。这些均充分表明了中国政府的温室气体减排决心,为中国应对气候变化、绿色低碳发展提供了方向指引、擘画了宏伟蓝图。

中国从基本国情和发展阶段的特征出发,大力推进生态文明建设,实施积极应对气候变化国家战略,把应对气候变化有机融入国家经济社会发展的中长期规划,坚持减缓和适应气候变化并重,通过法律、行政、技术、市场等多种手段,加快推进绿色低碳发展,主动控制温室气体排放,增强适应气候变化能力。中国政府也一如既往地信守应对全球气候变化的承诺,坚持共同但有区别的责任原则、公平原则和各自能力原则,全面落实国家适当减缓行动及强化应对气候变化行动的国家自主贡献,积极参与应对全球气候变化谈判,推动和引导建立公平合理、合作共赢的全球气候治理体系,深化气候变化多双边对话交流与务实合作,支持其他发展中国家加强应对气候变化能力建设,推动构建人类命运共同体。

党的十八大以来,在以习近平同志为核心的党中央坚强领导下,各地区各部门深入贯彻习近平生态文明思想,贯彻落实全国生态环境保护大会的部署和要求,实施积极应对气候变化的国家战略,积极落实"十三五"控制温室气体排放目标任务，在努力控制温室气体排放的同时主动开展适应行动，应对气候变化工作取得明显成效。通过调整产业结构、优化能源结构、节能提高能效、推进碳市场建设、提升适应气候变化能力、增加森林碳汇等一系列措施，中国单位国内生产总值二氧化碳排放（以下简称碳强度）持续下降，基本扭转二氧化碳排放快速增长局面。截至2019年底，碳强度比2015年下降18.2%，已提前完成"十三五"约束性目标任务；碳强度较2005年降低约48.1%，非化石能源占能源消费比重达15.3%，均已提前完成中国向国际社会承诺的2020年目标。经测算，相当于减少二氧化碳排放约56.2亿吨，减少二氧化硫约1192万吨、氮氧化物约1130万吨，应对气候变化和污染防治的协同作用初步显现。自2011年起，中国在北京、天津、上海、重庆、湖北、广东、深圳7个省（市）开展碳排放权交易试点，截至2020年8月底，试点碳市场累计配额现货成交量约4.06亿tCO_2e，成交额约92.8亿元。在总结借鉴试点经验的基础上，稳步推进全国碳市场建设。与此同时，积极建设性参与全球气候治理，始终坚持多边主义，为《巴黎协定》达成和生效发挥重要作用。积极开展气候变化南南合作，帮助其他发展中国家提升应对气候变化能力。中国在应对气候变化方面的努力获得了国际社会的积极肯定。

在看到成绩的同时，我们也要清醒地认识到，中国仍是最大的发展中国家，发展不平衡不充分问题突出，外部环境复杂严峻，目前在有关应对气候变化的认知水平、政策工具、手段措施、基础能力等方面还存在欠缺和短板，要实现新达峰目标与碳中和愿景，任务十分艰巨，需要付出艰苦卓绝的努力。因此，要善于将这一重大目标和愿景，转化为倒逼经济高质量发展和生态环境高水平保护，大力推动经济结构、能源结构、产业结构转型升级，推进新技术、新业态创新和涌现，推动构建绿色低碳循环发展的经济体系的新机遇。就生态环境领域而言，要进一步推进应对气候变化与生态环境保护工作统筹融合、协同增效；在减排目标、任务举措、管理制度、监测评价与执法监管等领域，加强温室气体与大气污染物排放协同控制，促进适应气候变化与生态系统保护修复等工作协同推进，努力开创生态文明建设和生态环境保护工作新局面。

1.1.1.2 雨养农业：气候变化影响最深远、最直接的生产部门

雨养农业（rainfed agriculture）是指在无灌溉的条件下，依靠天然降水为

水源进行农业生产的农业类型。雨养农业是世界上最大的农业生产部门，使用了全球75%以上的耕地，是全球超过一半人口的食物来源，也是全球发展中国家超过十亿贫困人口的生活来源。[①] 具体而言，在撒哈拉以南非洲，雨养农业所占耕地总面积的比重超过95%（例如，在埃塞俄比亚，比重为96.8%），提供了90%的粮食。[②] 在其他地区，雨养农业所占耕地总面积的比重为拉丁美洲90%、近东与北非地区75%、东亚地区65%以及南亚地区60%。在印度，雨养农业使用了超过60%的耕地，提供了44%的粮食、91%的杂粮、91%的豆类、80%的植物油料以及60%的棉花。[③] 在澳大利亚，雨养农业产出了绝大多数的粮食、饲料和纤维，其产值占全国农业产值的一半以上。[④] 在中国，雨养农业广泛分布于北部和西部的干旱半干旱地区，涉及超过1.5亿农民，所涵盖耕地面积约占全国耕地总面积的50%，小麦、玉米、谷子、高粱和豆类产量超过全国的50%。[⑤]

雨养农业是受气候变化影响最深远、最直接的生产部门。[⑥] 在雨养农业区，有效降水量往往直接决定了农业产量，而降水的季节性变化导致了产量的变动，农业生产对于气候变化尤为敏感。以中国为例，其雨养农业区多为大陆性干旱半干旱季风气候，局部地形与气候多变，不同季节水蚀风蚀交替出现，许多地区存在有明显的水土流失现象，生态环境较为脆弱，农业生产力水平长期偏低，农业生产极易受气候变化的影响。[⑦]

此外，发展中国家的雨养农业区域多为经济社会欠发达地区，当地农业

[①] Herrero, M., Thornton, P. K., Notenbaert, A. M., et al. Smart investments in sustainable food production: revisiting mixed crop-livestock systems [J]. Science, 2010, 327 (5967): 822–825.

[②] Sidibé, Y., Foudi, S., Pascual, U., Termansen, M. Adaptation to climate change in rainfed agriculture in the global south: soil biodiversity as natural insurance [J]. Ecological Economics, 2018, 146: 588–596.

[③] 参见https://krishi.icar.gov.in/jspui/bitstream/123456789/33282/1/GGSNR.pdf.

[④] ABS. Value of Agricultural Commodities Produced, Australia, 2009–10 [R]. Canberra: ABS, 2011.

[⑤] Eldoma, I. M., Li, M., Zhang, F., Li, F. M. Alternate or equal ridge-furrow pattern: Which is better for maize production in the rain-fed semi-arid Loess Plateau of China? [J]. Field Crops Research, 2016, 191: 131–138.

[⑥] Cholo, T. C., Fleskens, L., Sietz, D., Peerlings, J. Land fragmentation, climate change adaptation, and food security in the Gamo Highlands of Ethiopia [J]. Agricultural Economics, 2019, 50 (1): 39–49.

[⑦] 田贵良、贾琨颢、孙兴波等：《干旱事件影响下虚拟水期权契约的提出及其定价研究》，载《农业技术经济》2016年第9期。石晓丽、史文娇：《北方农牧交错带界线的变迁及其驱动力研究进展》，载《农业工程学报》2018年第20期。

基础设施薄弱，农户缺乏生产性资金，农业生产技术落后，贫困发生率偏高。① 在埃塞俄比亚东北部，1998～2000 年的持续干旱所导致的种植业和畜牧业损失约为每户 266 美元，超过了当地 75% 农户的年均现金收入。② 当前，与其他发展中国家的情况相类似，中国雨养农业对气候变化相对敏感且比较脆弱，面临着粮食保产以及农民增收所带来的巨大压力，同时也面临着气候变化所带来的一系列影响。1961～2010 年，作物生育期内平均温度升高使得冬小麦和玉米的单产出现下降。主要雨养农业区黄土高原地区是中国对于气候变化表现最为脆弱的地区，当地 90% 的耕地出现一种作物减产，55% 的耕地出现两种作物减产。③

即便在发达国家，气候变化对于雨养农业的影响也是显而易见的。在澳大利亚，2002～2010 年的千禧年干旱（millennium drought）导致雨养农业全要素生产率增速放缓，下降约 18%；农业生产总值在 2003 年和 2006 年分别下降了 24.8% 和 18%，农业出口和收入也相应地显著下降。④

现有研究显示，全球雨养农业区正经历着显著的气候变化。在澳大利亚，西南部雨养农业区的年平均温度自 1950 年以来每 10 年上升 0.1℃～0.3℃。年均降水量自 20 世纪 60 年代末开始呈显著下降趋势，平均每 10 年下降 5～30 毫米，自 2000 年以来下降的趋势更加明显（见图 1-13）。生产季节（秋季和冬季）降水的下降最为突出。预计到 2030 年，冬季和春季（五月至十月）降水量将进一步减少 10%，降水量下降幅度有可能超过 20 世纪后期⑤。

---

① 谭淑豪、谭文列婧、励汀郁等：《气候变化压力下牧民的社会脆弱性分析——基于内蒙古锡林郭勒盟 4 个牧业旗的调查》，载《中国农村经济》2016 年第 7 期。田云、张俊飚：《中国低碳农业发展的动态演进及收敛性研究》，载《干旱区资源与环境》2017 年第 3 期。王一超、郝海广、张惠远等：《农牧交错区农户生计分化及其对耕地利用的影响——以宁夏盐池县为例》，载《自然资源学报》2018 年第 2 期。

② Carter, M. R., Little, P. D., Mogues, T., Negatu. W. Shock, sensitivity and resilience: tracking the economic impacts of environmental disaster on assets in Ethiopia and Honduras [R]. Wisconsin, USA, 2004.

③ 生态环境部：《中华人民共和国气候变化第三次国家信息通报》，2018。

④ Sheng, Y., Xu, X. The productivity impact of climate change: Evidence from Australia's Millennium drought [J]. Economic Modelling, 2019, 76: 182-191.

⑤ Hughes, L., Steffen, W. The Critical Decade: Climate Change Science, Risks and Responses [R]. Climate Commission Secretariat, Australia, 2013.

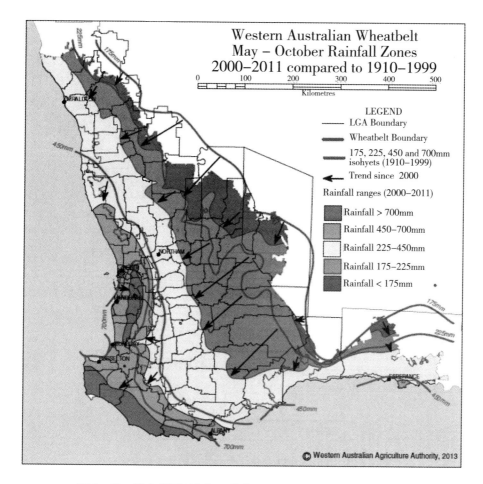

图 1-13　澳大利亚西南部雨养农业区生长季节降水量变化态势

注：图表示了生长季节（5~10 月）降水量 2000~2011 年水平与 1910~1999 年水平之间的差异。其中 175 毫米、225 毫米、450 毫米和 700 毫米等降水量线为 1910~1999 年平均水平；箭头体现了 2000 年以来的变化趋势；不同颜色体现了 2000~2011 年降水量水平。

资料来源：图片引自澳大利亚西澳大利亚州基础工业与地区发展部 https：//www.agric.wa.gov.au/drought-and-dry-seasons/evolution-drought-policy-western-australia? page=0%2C3。

在印度的雨养农业区，由于季风的变化，降水的季节变化更加极端，干旱与洪涝更加多发。1951~2016 年，印度干旱发生的频率和空间范围都有显著增加。在印度的一些主要雨养农业区，平均每 10 年遭遇至少 2 次干旱。受干旱影响的地区也每 10 年增加 1.3%。由于季风降水变动的增加以及温暖大气中水汽需求增加，预计到 21 世纪末，印度干旱的频率、范围以及强度都会增加，年气温中位值将上升约3℃。① 在非洲雨养农业区，由于气温升高，预计到2050年生长季节的长度将缩短5%~20%，在萨赫勒地区缩短的幅度可能会超过20%。②

在中国雨养农业区，一方面，气温呈现显著的上升趋势（见图1-14）。在西北雨养农业区，1961~2012年平均每10年气温上升约0.3℃，2001~2010年是近半个世纪以来最暖的时段，农业热量资源增加，冬小麦和玉米种

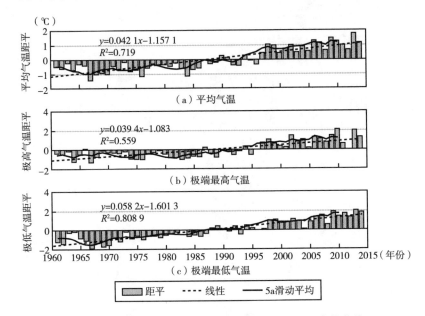

图1-14　1961~2014年中国西北地区气温距平变化曲线

资料来源：图片引自商沙沙等（2018）。商沙沙、廉丽姝、马婷等：《近54a中国西北地区气温和降水的时空变化特征》，载《干旱区研究》2018年第1期。

---

① Christensen, J. H., Hewitson, B., Busuioc, A., et al. Regional climate projections [R]. Contribution of Working Group I to the Fourth Assessment Report of the Intergovernmental Panel on Climate Change, ed. Solomon, S., Qin, D., Manning, M., et al. Cambridge, UK: Cambridge University Press, 2007. Government of India. Assessment of climate change over the Indian Region. A report of the Ministry of Earth Sciences (MoES), Government of India [R]. Springer Nature Singapore, Singapore, 2020.

② Cooper, P. J. M., Dimes, J., Rao, K. P. C., et al. Coping better with current climatic variability in the rain-fed farming systems of sub-Saharan Africa: An essential first step in adapting to future climate change? [J]. Agriculture Ecosystems & Environment, 2008, 126 (1-2): 24-35. Thornton, P. K., Steeg, J. V. D., Notenbaert, A., et al. The impacts of climate change on livestock and livestock systems in developing countries: A review of what we know and what we need to know [J]. Agricultural Systems, 2009, 101 (3): 113-127.

植北界北移。① 另一方面，雨养农业区的降水量变化明显（见图1-15）。在新疆北部、河西走廊、祁连山区等区域出现增加，但在陕西、宁夏、甘肃东部、青海东部、鄂尔多斯等区域减少趋势明显。② 西北雨养农业区东部春小麦和春玉米各生育阶段的干旱强度呈增加趋势，西部呈下降趋势；夏玉米干旱强度在陕西北部、宁夏和河西走廊呈增加趋势，其他地区多为降低趋势。③

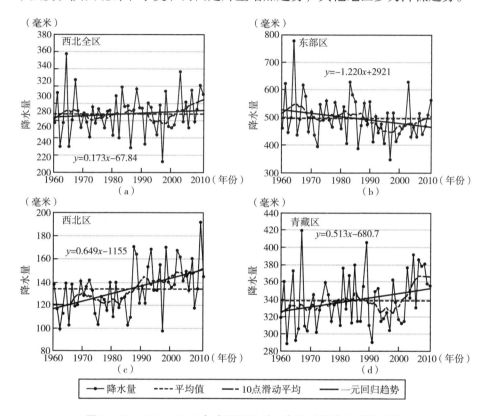

图1-15　1960~2011年中国西北地区年降水量年际变化趋势

资料来源：图片引自黄小燕等（2015）。黄小燕、李耀辉、冯建英等：《中国西北地区降水量及极端干旱气候变化特征》，载《生态学报》2015年第5期。

---

① 姚玉璧、杨金虎、肖国举等：《气候变暖对西北雨养农业及农业生态影响研究进展》，载《生态学杂志》2018年第7期。

② 刘玉洁、陈巧敏、葛全胜等：《气候变化背景下1981~2010中国小麦物候变化时空分异》，载《中国科学：地球科学》2018年第7期。马雅丽、郭建平、赵俊芳：《晋北农牧交错带作物气候生产潜力分布特征及其气候变化的响应》，载《生态学杂志》2019年第3期。

③ 何斌、刘志娟、杨晓光等：《气候变化背景下中国主要作物农业气象灾害时空分布特征（Ⅱ）：西北主要粮食作物干旱》，载《中国农业气象》2017年第1期。

气候变化增加了全球雨养农业区农业生产的不确定性，直接影响了农业生产布局和结构。因此，有必要弄清气候变化所带来的自然环境以及社会经济等方面变化对雨养农业生产的影响，并在此基础上探寻雨养农业对气候变化的适应性路径。这是合理预测未来可持续发展必不可少的工作，对于包括中国在内的世界各国特别是广大发展中国家应对气候变化、保障农业生产与粮食安全、减少农村贫困、促进区域环境经济协同发展有着重要意义，这也是当前学界较少涉及的一个新领域。本书正是基于这样的背景而开展的。

1.1.2 研究意义

雨养农业是世界上最大和最重要的农业生产部门。系统梳理雨养农业区气候变化的中长期趋势及其潜在变化幅度，分析气候变化对雨养农业生产的影响，基于微观证据和国际比较，探索在优化农户收益前提下雨养农业对气候变化的有效适应路径，具有重要的理论价值和实践意义。

首先，尝试构建一个将优化农户收益与雨养农业对气候变化有效适应有机融合的理论分析框架，是对农业可持续发展理论的深入和完善。考虑到生产经营实际情况和雨养农业的特征，各国雨养农业对气候变化的适应需要与优化农户收益相结合，二者缺一不可。本书在分析框架中，将优化农户收益与雨养农业对于气候变化有效适应进行有机融合，认为雨养农业对气候变化的适应必须在优化农户收益的前提下进行，这符合雨养农业生产实际情况和农业绿色可持续转型的要求，进一步补充和丰富农业经济学、环境经济学、生物经济学、自然地理学以及系统科学在农业可持续发展理论上的认识，具有重要的理论意义。

其次，探索研究雨养农业生产系统在气候变化条件下综合演化过程和路径的方法论，并提供一定的研究方案和方法参考。雨养农业区域的自然环境条件往往较为复杂，农户多综合经营种植业和畜牧业，生产规模差异较大。发达国家雨养农业多采用大农场规模化经营。发展中国家雨养农业生产规模较小，多为小农户经营模式，农业生产技术落后，贫困发生率偏高，对于气候变化的适应能力较为薄弱，且在适应过程中还存在着优化农户收益和确保粮食安全的客观需求，适应过程和路径更加复杂。现有关于气候变化条件下雨养农业生产的研究多建立在雨养农业系统的某一单一模块的基础上，对于各模块之间的相互关系以及共同作用涉及较少，与雨养农业实际存在较大差异。本书拟建立的雨养农业生产生物经济学模型，将雨养农业生产各模块之间的相互关系以及共同作用纳入其中，梳理多尺度和多要素之间的多维度相互作用，将自然地理学、农业经济学、气候变化经济学、农学和决策理论与方法等学科的理论和方法进行有机结合，基于国际研究比较，探索研究雨养农业生产系统在气候变化条件下综合演化过程和路径的方法论，为其提供了有参考价值的研究方案和方法。这对丰富和延伸气候变化适应研究的内容和深度，明晰雨养农业生产系统各部分的内在联系，规范农业可持续性适应研究范式有着重要的理论和实践意义。

最后，为雨养农业应对气候变化以及推动农业绿色发展与转型提供理论依据和有益参考。本书综合分析气候变化对世界主要雨养农业区的农业生产在农业生态要素、社会经济要素以及农户家庭等方面所造成的直接和间接影响，探索在农户收益优化的基础上，运用多种农业气候变化适应性措施组合及其综合影响，系统分析雨养农业生产对气候变化的有效适应路径。这将为合理预测未来可持续发展、保障农业生产、促进区域环境经济协同发展提供科学证据支撑，为雨养农业应对气候变化以及推动农业绿色发展与转型提供理论依据和有益参考，具有重要的实践意义。

1.2 研究目标

本书从雨养农业的具体特点和实际出发，基于已有的科学研究成果和实地观测数据，系统梳理雨养农业区气候变化的中长期趋势及其潜在变化幅度，分析气候变化对雨养农业生产的影响；通过国际比较，从微观的视角出发，分析雨养农业温室气体减排的成本有效性，探索在优化农户收益前提下雨养农业对于气候变化有效适应路径，为应对气候变化、保障农业生产与粮食安全、推动农业绿色发展与转型、促进区域环境经济协同发展提供理论依据和政策参考。具体而言，包括以下五个目标。

目标一：系统梳理雨养农业区气候变化的中长期趋势及其潜在变化幅度，构建雨养农业区气候变化情景。

目标二：分析比较各情景下气候变化对主要雨养农业区农业生产在农业生态要素、社会经济要素以及农户家庭等方面所造成的直接和间接影响，并揭示其内在机理。

目标三：建立一个基于农户、以优化农户收益为目标的，包含种植和畜牧部门气候变化适应性措施，综合考量生物、环境、管理、技术以及财务等方面条件的雨养农业生产生物经济学模型，将雨养农业生产各模块之间的相互关系以及共同作用纳入其中。

目标四：从适应行为、适应策略以及适应效果等方面探索雨养农业对气候变化的有效适应，系统分析雨养农业生产对气候变化的适应过程和有效适应路径。

目标五：运用生产者微观视角，系统地分析气候变化对雨养农业生产的影响，评价雨养农业对气候变化的环境效应和适应成本，探索在优化农户收益的前提下雨养农业对气候变化的有效适应路径的理论依据。

1.3 研究方法

本书以生物经济学、农业经济学、气候变化经济学、农学、自然地理学和决策理论与方法等学科的理论模型为基础，结合雨养农业生产的实际，通过文献研究、资料收集与分析、数学规划、生物物理学分析、情景分析、生产力分析、计量分析、案例分析、国际比较相结合的方法来展开研究。

（1）文献研究。本书拟通过查阅分析实地观测数据、文献数据和相关研究文献成果的方法，获得所需数据资料，主要包括：澳大利亚西南部地区和中国黄土高原地区农业气象试验站所记录的植株生长、田间管理和土壤参数等长期观测数据、澳大利亚气象局和中国气象局地面气象观测资料、《中国土种志》《中华人民共和国土壤图》以及澳大利亚土壤数据库有关数据、澳大利亚西澳大利亚州以及中国陕西、山西、甘肃、宁夏、内蒙古、新疆等省区的畜牧志、农业志和统计年鉴的有关数据；借助爱思唯尔（Elsevier）、施普林格（Springer Link）、科学网络引文数据库（Web of Science）、谷歌学术（Google Scholar）和中国知网、万方、维普等文献全文数据库中近年来涉及地区有关研究的相关成果。在此基础之上，结合研究所需，对收集到的数据进行综合分析与筛选。

（2）数学规划。考虑到已有研究经验和雨养农业生产的实际，本书拟利用数学规划中的动态混合整数规划法，对扎莫等（Thamo et al.）所提出的雨养混合农业农户生产静态优化分析框架进行拓展，建立一个基于雨养农业生产系统整体，综合考量生物、环境、管理、技术以及财务等各方面因素，基于农户的动态优化生物经济学模型。① 扎莫等（2017）所提出的雨养混合农业农户生产静态分析框架及其具体改进版，已被运用到对澳大利亚、埃塞俄比亚、南非和中亚地区主要雨养农业区农业生产的一系列分析当中，相关模型的模拟结果良好。②

本书所使用的模型涵盖雨养农业区主要的农业活动，包括每一个农地管理单元上的作物——牧草轮作、牲畜饲料的供给与使用、化肥的使用、现金流的记录、农机成本与间接开销等。模型基于雨养农业区农户的动态农地利用和生产过程，并且将农地利用的动态影响（如作物轮作对于未来产量预估的影响）包含在内。模型的目标函数是对将农户收入扣除变动成本以及间接费用后所得到的毛利进行最大化。模型的最优解描述的是一个基于农户可利用资源产生最大化毛利润的一系列农业生产活动序列的集合。模型包括一系列的约束条件，可分为资源约束（如农机总动力）、逻辑约束、技术约束（如羊群饲料需求）等。

该模型中的代表性农作物类型包括小麦、玉米、油菜、豆类和牧草，代表性的牲畜是羊。农作物、活羊以及羊毛是农场的主要产品。模型假设农民依据农产品的预期价格、农业经营成本以及面临的环境条件（如土壤类型和生长季节降水量等）来做出农业生产决策。模型中雨养农业生产的气候变化适应措施包括保护性耕作、轮作、改善肥料施用、休耕以及改善畜群经营。

在本书开展的研究中，将主要雨养农业区的气候环境数据、雨养农业生产数据以及其他经济数据分别导入模型中进行参数调整与验证工作，以不断提高改进模型模拟精度。模型的编写和操作利用 GAMS（General Algebraic Modelling System，ver 24.8.2）软件进行。

（3）生物物理学分析。本书拟建立的雨养农业生产生物经济学模型中作

---

① Thamo, T., Addai, D., Pannell, D. J., et al. Climate change impacts and farm-level adaptation: Economic analysis of a mixed cropping-livestock system [J]. Agricultural Systems, 2017, 150: 99 – 108.

② Ghahramani, A., Bowran, D. Transformative and systemic climate change adaptations in mixed crop-livestock farming systems [J]. Agricultural Systems, 2018, 164: 236 – 251. Tang, K., Hailu, A., Kragt, M. E., Ma, C. The response of broadacre mixed crop-livestock farmers to agricultural greenhouse gas abatement incentives [J]. Agricultural Systems, 2018, 160: 11 – 20. Wossen, T., Berger, T., Haile, M. G., Troost, C. Impacts of climate variability and food price volatility on household income and food security of farm households in East and West Africa [J]. Agricultural Systems, 2018, 163: 7 – 15. Tang, K., Hailu, A. Smallholder farms' adaptation to the impacts of climate change: Evidence from China's Loess Plateau [J]. Land Use Policy, 2020, 91, 104353.

物与牲畜的生产模块主要从 APSIM（Agricultural Production Systems Simulator, ver 7.5）和 GrassGro（ver 3.2.6）这两个生物物理学模型中提取。这两种模型已被应用在包括中国、美国、澳大利亚、埃塞俄比亚、南非、中亚地区和环地中海地区在内的全球不同的雨养农业区，相关研究进行了不少参数调整和验证作业，模型模拟结果良好。① 本书中 APSIM 和 GrassGro 模型的具体参数调整与验证工作将结合澳大利亚西南部地区和中国黄土高原地区实地观测数据和文献数据进行。

（4）情景分析。受限于各种气候模型本身的局限，以及对于未来全球温室气体排放量的不确定，目前学界对雨养农业区气候变化潜在变化幅度的判断难以达成共识，仍然存在较大的不确定性。为了有效反映这种不确定性，本书将各气候变化关键因子的变化情景设定在一个较大的范围。类似的情景要素分析法近年来已被不少的气候变化研究所采用。② 本书将不同的气候变化关键因子进行组合，形成一系列的雨养农业区气候变化情景。各气候变化关键因子变化幅度的选择，既基于已有研究结果和气象局地面气象观测资料，同时也尽可能地将未来气候变化的潜在幅度包含在内，这将有助于进一步探究雨养农业对于气候变化的敏感性。此外，考虑到数据的可得性，由合理周期内的主要雨养农业区各气候变化关键因子的平均值，构成气候基本情景。

依据已确定的雨养农业区气候变化情景，利用模型对澳大利亚西南部地

---

① Kragt, M. E., Pannell, D. J., Robertson, M. J., Thamo, T. Assessing costs of soil carbon sequestration by crop-livestock farmers in Western Australia [J]. Agricultural Systems, 2012, 112: 27-37. Thamo, T., Addai, D., Pannell, D. J., et al. Climate change impacts and farm-level adaptation: Economic analysis of a mixed cropping-livestock system [J]. Agricultural Systems, 2017, 150: 99-108. Doran-Browne, N., Wootton, M., Taylor, C., Eckard, R. Offsets required to reduce the carbon balance of sheep and beef farms through carbon sequestration in trees and soils [J]. Animal Production Science, 2018, 58 (9): 1648-1655. Robertson, S. M., Friend, M. A. Performance of sheep systems grazing perennial pastures. 4. Simulated seasonal variation and long-term production [J]. Animal Production Science, 2020, 60 (3): 423-435. 戴彤、王靖、赫迪等：《基于 APSIM 模型的气候变化对西南春玉米产量影响研究》，载《资源科学》2016 年第 1 期。

② Meier, E. A., Thorburn, P. J., Kragt, M. E., et al. Greenhouse gas abatement on southern Australian grains farms: Biophysical potential and financial impacts [J]. Agricultural Systems, 2017, 155: 147-157. Thamo, T., Addai, D., Pannell, D. J., et al. Climate change impacts and farm-level adaptation: Economic analysis of a mixed cropping-livestock system [J]. Agricultural Systems, 2017, 150: 99-108. Tang, K., Hailu, A., Kragt, M. E., Ma, C. The response of broadacre mixed crop-livestock farmers to agricultural greenhouse gas abatement incentives [J]. Agricultural Systems, 2018, 160: 11-20. Tang, K., Hailu, A. Smallholder farms' adaptation to the impacts of climate change: Evidence from China's Loess Plateau [J]. Land Use Policy, 2020, 91, 104353.

区和中国黄土高原地区的雨养农业在不同气候变化情景下，采取气候变化适应策略组合后的农业产量、种—畜结构、粮食作物与经济作物比例、农地分配、农业温室气体排放以及农业利润等方面的结果进行模拟。所得出的结果将与气候基本情景下的模拟结果、实地观测数据以及文献数据进行比较，分析气候变化对主要雨养农业区农业生产在农业生态要素、社会经济要素以及农户家庭等方面所造成的直接和间接影响，并在此基础上，比较分析并概括出各气候变化情景下，在确保农户毛利最大化前提下，能够实现的最优适应策略，以及具体适应路径。

（5）生产力分析。考虑到雨养农业生产的实际，本书采用截尾正态随机前沿生产函数的二次形式来估计作物的产量。这将允许在生产过程中存在无效，即产量可以低于既定生产技术条件下的最优值，以及过量农业投入（如发生过量施肥和降水过多等）导致产量递减的情况发生。作物产量由施肥量、降水量、生产随机前沿的噪声项和技术无效来决定，其中，技术无效被假设为是一个关于土地和农户特征的函数，该函数包含一个时间趋势变量来分析生产技术进步的效果。

1.4 创 新 点

本书拟将生物经济学、农业经济学、气候变化经济学、农学、自然地理学和决策理论与方法等学科进行有机融合，系统梳理雨养农业区气候变化的中长期趋势及其潜在变化幅度，基于微观生产者视角和国际比较，分析气候变化对雨养农业生产的影响，探索在优化农户收益前提下雨养农业对气候变化的有效适应路径，主要的特色与创新之处具体如下。

（1）研究视角创新。本书在系统梳理已有文献基础上，基于生物经济学和全农场的视角，通过国际比较，进行了雨养农业微观层次上的系统研究。这克服了已有文献从单一适应措施和单一适应政策出发研究气候变化条件下雨养农业生产的局限，可以较为有效地避免因研究系统选择过小带来的以偏概全的缺陷。

（2）在分析框架中将优化农户收益与雨养农业对气候变化的有效适应进行有机融合。本书认为，雨养农业对气候变化的适应需要与优化农户受益相结合，二者缺一不可。这是由农业生产实际情况和雨养农业的特征所决定的。

已有研究大多研究雨养农业部门某一部分对气候变化的适应，缺乏整体性和系统性，其分析框架与实际情况存在较大差异，所得出的结论存在较大的局限性。本书在分析框架中，将优化农户收益与雨养农业对气候变化的有效适应进行有机融合，认为雨养农业对气候变化的适应必须在优化农户收益的前提下进行，这符合雨养农业生产实际情况和农业绿色可持续转型的要求，能够提高研究结论的针对性和有效性，故具有较强的理论价值和应用价值。

（3）建立一个基于农户，以优化农户收益为目的，包含种植和畜牧部门气候变化适应性措施，综合考量生物、环境、管理、技术以及财务等方面条件的雨养农业生产生物经济学模型。该模型是一个基于雨养农业生产系统整体，综合考量各方面因素、基于农户的动态优化农业模型。该模型基于雨养农业区农户的动态农地利用和生产过程，涵盖雨养农业区种植和畜牧部门主要的农业活动，且将农地利用的动态影响包含在内。该模型将有效弥补已有研究的不足，为系统分析和理解雨养农业在气候变化条件下各模块之间的相互关系及其共同作用、比较多种农业气候变化适应性措施、分析各种措施组合的综合影响提供一个实用的分析工具。

第 2 章 文献综述

2.1 气候变化条件下雨养农业的相关生物物理学基础[*]

地球大气对于短波或可见太阳辐射的吸收力较弱,但对长波辐射的吸收力较强。因而太阳短波辐射可以穿透大气到达地面,但地面增温后释放出的长波辐射却被大气吸收,产生地表和低层大气温度增加的效应,即温室效应(见图 2-1)。[①] 能够吸收地面反射的太阳辐射的气体即温室气体。减少人类活动所产生的温室气体,被学界普遍认为是应对气候变化的最重要的途径。

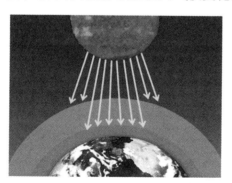

图 2-1 温室效应示意图

固碳是指将碳存储于陆地上的植被或土壤中,从而减少大气中碳含量的

---

[*] 该节部分内容参考唐凯:《基于生物经济学的澳大利亚农业温室气体减排潜能分析》,人民出版社 2018 年版。

[①] 引自中国香港天文台,http://www.hko.gov.hk/wxinfo/climat/greenhs/c_grnhse.htm。

过程。① 农业部门主要产生三种温室气体：二氧化碳、氧化亚氮和甲烷。其数量均可以二氧化碳当量（CO_2e）来度量。

通过（永久地）改变土地利用和变更现有土地利用中的管理行为，土壤能从温室气体源转变为温室气体吸收汇（GHG sink）。② 植物通过光合作用将二氧化碳转化为有机碳。有研究表明，植物干物质中约有四成的碳来自光合作用的固定。③ 其中的一些碳经由地上以及地下的分解转变为土壤有机碳（soil organic carbon），因而从大气中被移除，被固定于土壤中。④

大气中不断上升的二氧化碳水平促使众多学者关注土壤有机碳这一潜在的温室气体吸收汇。⑤ 一些学者估计全球土壤存储了 1.5×10^{12} 吨的土壤有机碳，约是大气中碳含量的2倍，以及全球陆地生物量（biomass）中碳含量的3倍。⑥ 在澳大利亚，每年农耕地潜在固碳量预计为6800万 tCO_2e。⑦ 大量的研究表明，改变农业土地利用及生产操作方式可以提高农用土地中碳的存储量。⑧

---

① Bernoux, M., Cerri, C. C., Volkoff, B., et al. Gaz à effet de serre et stockage du carbone par les sols, inventaire au niveau du Brésil [J]. Cahiers Agriculrures, 2005, 14: 96 – 100. Pendell, D. L., Williams, J. R., Boyles, S. B., et al. Soil carbon sequestration strategies with alternative tillage and nitrogen sources under risk [J]. Applied Economic Perspectives and Policy, 2007, 29 (2): 247 – 268.

② Lal, R. Soil management and restoration for C sequestration to mitigate the accelerated greenhouse effect [J]. Progress in Environmental Science, 1999, 1 (4): 307 – 326.

③ 闵九康：《全球气候变化和低碳农业研究》，气象出版社2011年版。

④ Fynn, A. J., Alvarez, P., Brown, J. R., et al. Soil carbon sequestration in US rangelands: issues paper for protocol development [R]. Environmental Defense Fund, 2009.

⑤ Oenema, O., Witzke, H. P., Klimont, Z., et al. Integrated assessment of promising measures to decrease nitrogen losses from agriculture in EU – 27 [J]. Agriculture, Ecosystems & Environment, 2009, 133 (3): 280 – 288. Sanderman, J., Farquharson, R., Baldock, J. Soil carbon sequestration potential: a review for Australian agriculture [R]. CSIRO Sustainable Agriculture National Research Flagship, 2010.

⑥ Lal, R. Soil carbon sequestration impacts on global climate change and food security [J]. Science, 2004, 304: 1623. Schlesinger, W. H., Andrews, J. A., Soil respiration and the global carbon cycle [J]. Biogeochemistry, 2000, 48 (1): 7 – 20. Bernoux, M., Cerri, C. C., Cerri, C. E. P, et al. Cropping systems, carbon sequestration and erosion in Brazil, a review [J]. Agronomy for Sustainable Development, 2006, 26: 1 – 8.

⑦ Garnaut, R. The Garnaut Climate Change Review [M]. Cambridge: Cambridge University Press, 2008.

⑧ Campbell, C. A., Janzen, H. H., Paustian, K., et al. Carbon storage in soils of the North American Great Plains: effect of cropping frequency [J]. Agronomy Journal, 2005, 97 (2): 349 – 363. Desjardins, R. L., Kulshreshtha, S. N., Junkins, B., et al. Canadian greenhouse gas mitigation options in agriculture [J]. Nutrient Cycling in Agroecosystems, 2001, 60 (1 – 3): 317 – 326. Grace, P. R., Antle, J., Ogle, S., et al. Soil carbon sequestration rates and associated economic costs for farming systems of south-eastern Australia [J]. Australian Journal of Soil Research, 2010, 48: 1 – 10. Kragt, M. E., Pannell, D. J., Robertson, M. J., et al. Assessing costs of soil carbon sequestration by crop-livestock farmers in Western Australia [J]. Agricultural Systems, 2012, 112: 27 – 37. Sanderman, J., Farquharson, R., Baldock, J. Soil carbon sequestration potential: a review for Australian agriculture [R]. CSIRO Sustainable Agriculture National Research Flagship, 2010.

在中国，2010年土地利用、土地利用变化和林业吸收二氧化碳10.30亿吨，排放甲烷174.0万吨，净吸收量为9.93亿tCO$_2$e。林地、农地、草地、湿地分别吸收7.79亿、0.66亿、0.45亿、0.45亿吨二氧化碳。① 基于中国农业土地利用管理下的土壤碳汇潜力估算，尤其是在推行优化管理措施下，未来50年中国农业土壤固碳减排潜力约为87~393 TgC·a^{-1}，相当于抵消中国工业温室气体排放总量的11%~52%。其中，有机肥应用、秸秆还田、保护性耕作措施下的土壤固碳份额分别为6%~11%、15%~21%和2.8%~11%，农田管理措施固碳份额总计约占30%~36%（相当于抵消工业温室气体排放3.4%~19%）；侵蚀管理/退化土壤恢复/荒漠化治理等措施下的土壤固碳份额约占64%~70%（相当于抵消工业温室气体排放8%~33%）。②

除此之外，改变包括农业管理活动在内的农业土地利用模式还可以减少排放到大气中的碳的数量。③ 改变农业土地利用模式可以影响土壤的有机质水平、土壤质地、pH值、水分含量、温度等因素，从而增加土壤的固碳作用，最终减少农地碳排放量。许多研究的结果都支持这一结论。④ 可以说，土壤是二氧化碳的净储库。⑤

氧化亚氮是一种强温室气体，其100年全球暖化潜能值（global warming potential，GWP）⑥ 为298，即在100年中氧化亚氮对全球暖化的影响是同等质量的二氧化碳的298倍。农业中的氧化亚氮主要产生于土壤、肥料、牲畜

---

① 生态环境部：《中华人民共和国气候变化第三次国家信息通报》，2018。
② 王小彬、武雪萍、赵全胜等：《中国农业土地利用管理对土壤固碳减排潜力的影响》，载《中国农业科学》2011年第11期。
③ Kragt, M. E., Pannell, D. J., Robertson, M. J., et al. Assessing costs of soil carbon sequestration by crop-livestock farmers in Western Australia [J]. Agricultural Systems, 2012, 112: 27-37.
④ Sanderman, J., Farquharson, R., Baldock, J. Soil carbon sequestration potential: a review for Australian agriculture [R]. CSIRO Sustainable Agriculture National Research Flagship, 2010. Desjardins, R. L., Kulshreshtha, S. N., Junkins, B., et al. Canadian greenhouse gas mitigation options in agriculture [J]. Nutrient Cycling in Agroecosystems, 2001, 60: 317-326. Grace, P. R., Antle, J., Ogle, S., et al. Soil carbon sequestration rates and associated economic costs for farming systems of south-eastern Australia [J]. Australian Journal of Soil Research, 2010, 48: 1-10.
⑤ 闵九康：《全球气候变化和低碳农业研究》，气象出版社2011年版。
⑥ 全球暖化潜能值就是比较不同温室气体排放在100年的时期内对于气候的影响。每种温室气体的全球暖化潜能值取决于两个量，即一定量的气体导致变暖的程度以及它持续存在大气中的时间。例如，若温室气体A与B在大气中所存在的时间相同，但A所引起的变暖效应是B的2倍，则A的全球暖化潜能值便是B的2倍。而若A与B所引起的变暖效应相同，但A存在于大气中的事件是B的2倍，则A的全球暖化潜能值也是B的2倍。目前一般将CO$_2$的全球暖化潜能值规定为1。

栏舍以及牧草沉积。① 土壤中的硝酸盐容易因为过量的水流而被浸滤。约有1%的被浸滤氮最终会转化为氧化亚氮。因此，强降雨与过度灌溉能增加农业的氧化亚氮排放量。此外，氮是农作物生产所必需的物质。农民通常会施放含氮的化肥来提升产量。这容易使得土壤中包含过量的氮，增加被浸滤氮以及被排放的氮的数量。② 有证据显示，肥料是在土壤施肥条件下最主要的氧化亚氮的来源。③

　　甲烷的100年全球暖化潜能值为25，意味着其在100年中对全球暖化的影响是同等质量的二氧化碳的25倍。甲烷主要产生于厌氧条件下有机质的分解。④ 在农业领域，反刍牲畜的肠道发酵以及粪便是主要的甲烷排放源。⑤ 有研究显示，每年反刍牲畜排放出0.8亿～1.15亿吨甲烷，约占全球全年甲烷排放总量的1/5。⑥ 反刍牲畜体内的产甲烷菌引起肠道发酵，这一过程受牲畜种类、载畜率以及饲料的影响。⑦ 弗莱鲍尔等（Freibauer et al.）发现，储存中的粪便会产生甲烷，其中大部分来自液态粪便。⑧ 甲烷的排放率受到温度变化、（粪便）储藏方法以及粪便来源（牲畜种类）等因素

---

① Jackson, J., Choudrie, S., Thistlethwaite, G., et al. UK Greenhouse Gas Inventory, 1990 to 2007: Annual Report for submission under the Framework Convention on Climate Change [R]. UK Department of Energy and Climate Change, 2009. Abberton, M. T., Marshall, A. H., Humphreys, M. W., et al. Genetic Improvement of Forage Species to Reduce the Environmental Impact of Temperate Livestock Grazing Systems [J]. Advances in Agronomy, 2008, 98: 311 – 355. Smith, P., Martino, D., Cai, Z., et al. Greenhouse gas mitigation in agriculture [J]. Philosophical Transactions Royal Society of London Biological Sciences, 2008, 363: 789 – 813.

② Machefert, S. E., Dise, N. B., Goulding, KWT et al. Nitrous oxide emission from a range of land uses across Europe [J]. Hydrology and Earth System Sciences, 2002, 6: 325 – 337.

③⑥ 闵九康：《全球气候变化和低碳农业研究》，气象出版社2011年版。

④ Smith, P, Martino, D, Cai, Z, et al. Greenhouse gas mitigation in agriculture [J]. Philosophical Transactions Royal Society of London Biological Sciences, 2008, 363: 789 – 813.

⑤ IPCC. 2006 Guidelines for National Greenhouse Gas Inventories. Intergovernmental Panel on Climate Change [R]. National Greenhouse Gas Inventories Programme, IPCC, 2006.

⑦ Moorby, J. M., Chadwick, D. R., Scholefield, D., et al. A review of best practice for reducing greenhouse gases [R]. Defra project report AC0206, 2007. Smith, P, Martino, D, Cai, Z, et al. Greenhouse gas mitigation in agriculture [J]. Philosophical Transactions Royal Society of London Biological Sciences, 2008, 363: 789 – 813. Freibauer, A. Regionalised inventory of biogenic greenhouse gas emissions from European agriculture [J]. European Journal of Agronomy, 2003, 19: 135 – 160. 董红敏、李玉娥、陶秀萍等：《中国农业源温室气体排放与减排技术对策》，载《农业工程学报》2008年第10期。

⑧ Freibauer, A. Regionalised inventory of biogenic greenhouse gas emissions from European agriculture [J]. European Journal of Agronomy, 2003, 19: 135 – 160.

的影响。① 这些影响因素均可通过人为干预措施来控制。其他的甲烷排放源还包括水稻田、生物质燃烧、城市废物以及矿物燃料的运输和生产等。②

总体而言，农业温室气体排放源包括燃料消耗（发生于农药喷洒、灌溉、耕作、牲畜栏舍与暖房的加热与照明、作物干燥等过程中），土壤排放，牲畜（肠道发酵、粪便及其储存）、土壤和植物生物质中存储的碳。③ 在澳大利亚，牲畜养殖与作物生产是农业中最主要的排放源（见图2-2）。在澳大利亚主要农业产区之一的西澳大利亚州（简称西澳），近八成农业温室气体由牲畜养殖与作物生产所产生。2009~2010年度牲畜（主要是绵羊和牛）肠道发酵排放了占全州农业排放量53.49%的温室气体，同期农业土壤排放所占份额为21.45%。④

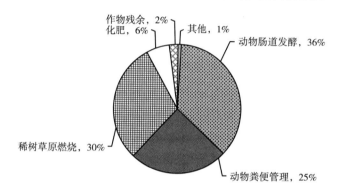

图2-2　2014年澳大利亚农业温室气体排放源构成

资料来源：联合国粮食与农业组织，《2016粮食及农业状况：气候变化、农业与粮食安全》，2016年。

在中国，牲畜养殖与作物生产同样是农业中最主要的排放源。2014年，中国大陆地区农业活动温室气体排放量约为7.08亿 tCO_2e，其中，动物肠道发酵排放量为2.04亿 tCO_2e，占28.8%；动物粪便管理排放量为1.94亿

---

① Monteny, G. J., Bannink, A., Chadwick, D. Greenhouse gas abatement strategies for animal husbandry [J]. Agriculture, Ecosystems & Environment, 2006, 112 (2): 163 - 170. Sommer, S. G., Petersen, S. O., Sørensen, P, et al. Methane and carbon dioxide emissions and nitrogen turnover during liquid manure storage [J]. Nutrient Cycling in Agroecosystems, 2007, 78: 27 - 36.
② 闵九康：《全球气候变化和低碳农业研究》，气象出版社2011年版。
③ Lewis, K. A., Tzilivakis, J., Green, A., et al. The climate change mitigation potential of an EU farm: towards a farm-based integrated assessment [R]. University of Hertfordshire, UK, 2010.
④ Australian Department of Climate Change and Energy Efficiency. Australian national greenhouse accounts: state and territory greenhouse gas inventories [R]. Australian Department of Climate Change and Energy Efficiency, 2012.

tCO$_2$e，占 27.5%；化肥排放量为 1.54 亿 tCO$_2$e，占 21.8%；水稻种植排放量为 1.13 亿 tCO$_2$e，占 15.9%；作物残余排放量为 0.36 亿 tCO$_2$e，占 5.1%；农业废弃物田间燃烧排放量为 0.05 亿 tCO$_2$e，占 0.7%（见图 2-3）。①

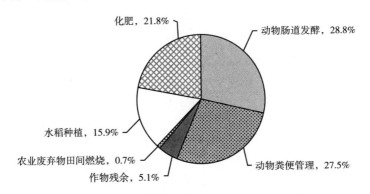

图 2-3　2014 年中国农业温室气体排放源构成

资料来源：联合国粮食与农业组织：《2016 粮食及农业状况：气候变化、农业与粮食安全》，2016 年。

对气候变化的应对指的是通过采取适当的调整与改变来减少气候变化所带来的负面影响或充分利用气候变化所带来的有利机会。其具体包括三个潜在目标，即减少损害发生的风险、培养应对无法避免损害的能力以及充分利用新的机会。在雨养农业部门中，结合当地的气候、土壤、水文等自然条件，农业经营者可以考虑通过以下措施来应对气候变化。

（1）保护性耕作（conservation tillage）（包括少耕或免耕②）。传统的耕作方法在一季作物收获后、下一季作物播种前都要进行翻地，造成土壤有机碳的暴露，刺激了微生物的活性。这些微生物可以将有机物中所包含的养分进行吸收，并将其分解，在此过程中会产生二氧化碳并释放到大气中。保护性耕作由于减少了对表层土壤的扰动，使得作物的根系活力增加，根系的生物量提高，土壤的呼吸强度降低，耕作土壤理化性状得以改善，土壤团聚体数量增加，不但能减少土壤中碳的损失，还能增加土壤的固碳率。③ 同时，保

---

① 联合国粮食与农业组织：《2016 粮食及农业状况：气候变化、农业与粮食安全》，2016 年。
② 少耕是指在取消铧式犁耕翻的基础上，仅为了保证播种、除草等农事保留少量扰动土壤的作业，减少耕作次数以及强度；免耕具体是指作物播种前不用犁、耙等农具整理土地，而选择在茬地上直接播种，而后在作物生长期间不使用农具进行土壤管理的耕作方法。
③ 张四海、曹志平、张国：《保护性耕作对农田土壤有机碳库的影响》，载《生态环境学报》2012 年第 2 期。

护性耕作减少了耗能的农机深耕作业与翻耕的需要，减少了燃料的使用，亦可降低温室气体的排放。目前澳大利亚的雨养农业部门已基本上采用了保护性耕作方式，在西澳大利亚州和南澳大利亚州保护性耕作采用率已分别达到88%和70%。在众多发展中国家，雨养农业部门保护性耕作的采用率还有待提高。近年来，在非洲的布基纳法索、塞内加尔、尼日尔和肯尼亚，以及亚洲的中国、印度，保护性耕作开始被越来越多的农民所采用。

（2）降低休耕频率（reducing fallowing frequency）。采用持续性种植，降低休耕频率，在增加作物产量的同时，增加了土壤作物残茬和根的有机质输入。虽然与此同时其微生物的降解率也相应提高，但总体上降解小于输入，土壤有机质含量得到提高，土壤有机质分解速率相对减缓，进而使得土壤中有机碳的含量提高。①

（3）轮种（rotational cropping）。采用谷类和豆科作物/豆科牧草轮种的方式，可以提高土壤中的有机质含量，增加土壤中氮的含量，减少谷类生长过程中氮肥的施用量，从而可以减少温室气体的排放；同时有机质的增加使得植物的营养再循环能力增强，有助于减少土壤中养分的流失，提高固碳率以及作物的耐旱能力。②

（4）混种（mixed cropping）。混种指的是在同一块农田里种植两种及以上的作物。在非洲的坦桑尼亚，农民普遍在农田里将谷类作物（玉米和高粱）、豆科作物以及花生种植在一起。混种的优势包括不同的成熟期（如玉米和豆科作物）、提高耐旱能力（玉米和高粱）、减少施肥需求（谷类和豆科作物）以及满足农民和更多的需求（如玉米用作家庭食物，而向日葵用来出售）等。研究发现，除了喀麦隆和南非以外的撒哈拉以南非洲各国，种植同一作物的不同品种被认为是最有效的气候变化应对措施之一。③

（5）退耕还林（afforestation or reforestation of agricultural lands）。树木是将空气中的碳以生物量的形式进行存储的最有效的载体。④ 树木中所含有的碳在转化之前都是以二氧化碳的形式在大气中存在，被树木以光合作用

---

① 杨景成、韩兴国、黄建辉等：《土壤有机质对农田管理措施的动态响应》，载《生态学报》2003年第4期。

② 张四海、曹志平、张国：《保护性耕作对农田土壤有机碳库的影响》，载《生态环境学报》2012年第2期。

③ Akinnagbe, O. M., Irohibe, I. J. Agricultural adaptation strategies to climate change impacts in Africa: a review [J]. Bangladesh Journal of Agricultural Research, 39 (3): 407-418.

④ 有研究表明，一颗道格拉斯冷杉在100年的生长过程中最多可吸收800吨CO_2。

的形式进行转化。退耕还林在增加对于大气中碳吸收量的同时，可以增加土壤中有机碳和其他有机质的含量，减少水土流失，是行之有效的农业气候应对措施。①

（6）增加秸秆覆盖（increasing retention of crop residues）。增加秸秆覆盖使得土壤的团粒结构得到了改善，土壤中养分含量得以提高，促使作物的生长，从而间接增加土壤的碳输入；同时秸秆覆盖可以有效减少土壤的风蚀、水蚀，降低无效蒸发量，有利于减少土壤中有机碳以及水分的流失。② 增加秸秆覆盖可以恢复土地肥力，从而可以减少作物在生长过程中所需施用的化肥，进而减少温室气体的排放。

（7）改变施肥方法（changing fertilisation practices）。如果按照适当的比例在合适的时间施用化肥，可以增加土壤中有机质的含量，有助于土壤中有机碳含量的提高；同时采用免耕与施用有机肥以及有机肥与化肥搭用的方式，可以提高土壤中碳氮含量，从而减少温室气体的排放。③

（8）作物多样化（crop diversification）。中长期而言，种植更多经济价值高的作物是雨养农业应对气候变化的一条可行途径。总体来说，对于非灌溉以及灌溉地区而言，作物多样性都是一项值得优先考虑的气候变化应对措施。在南非，许多农民从传统的畜牧农业转变为特种养殖农业（game farming）。④ 在苏丹西部的科尔多凡和达尔富尔地区，农民将粮食作物替换成经济作物，并开始种植适应能力更强的作物。⑤ 在坦桑尼亚，农民将种植作物种类多样化作为分散农业生产风险的有效方式。⑥ 作物多样化在一定程度上可以作为应对降水变率的保险。

---

① 张四海、曹志平、张国：《保护性耕作对农田土壤有机碳库的影响》，载《生态环境学报》2012 年第 2 期。

② Australian Department of Climate Change and Energy Efficiency. Australian national greenhouse accounts: state and territory greenhouse gas inventories [R]. Australian Department of Climate Change and Energy Efficiency, 2012.

③ 周莉、李保国、周广胜：《土壤有机碳的主导影响因子及其研究进展》，载《地球科学进展》2005 年第 1 期。

④ Ziervogel, G., Cartwright, A., Tas, A., et al. Climate change and adaptation in African agriculture [R]. Stockholm Environment Institute, 2008.

⑤ Akinnagbe, O. M., Irohibe, I. J. Agricultural adaptation strategies to climate change impacts in Africa: a review [J]. Bangladesh Journal of Agricultural Research, 39 (3): 407 – 418.

⑥ Adger, W. N., Huq, S., Brown, K., et al. Adaptation to climate change in the developing world [J]. Progress in Development Studies, 2014, 3 (3): 179 – 195.

(9) 改变牲畜和放牧管理方式 (changing livestock and grazing management)。反刍动物饲料的精粗比与其瘤胃甲烷产量之间存在紧密的联系。适量提高饲料中精料的比重利于降低动物体内乙酸的产生,从而减少甲烷的排放。同时,对于动物粪便进行表面覆盖可以减少甲烷的产生;避免粪便处于干湿交替状态还可以抑制氧化亚氮的生成与排放。① 此外,应对气候变化的具体措施还包括:识别当地适应气候压力以及饲料来源的牲畜品种并进一步强化其适应能力;通过与耐旱且疾病抵抗能力较强的品种进行杂交繁育以改良当地牲畜品种。②

在非洲的萨赫勒地区,牧民所采取的气候变化适应具体措施包括:在干旱期间提供应急饲料、增加畜群品种以应对极端气候事件以及在干旱期间屠宰体弱的牲畜作为食物。在干旱期间,考虑到可获得饲料的减少,牧民和农牧民会从养牛转变为饲养绵羊和山羊。萨赫勒地区的牧民通过在干旱的北部地区与较为湿润的南部地区之间的循环游牧迁徙来减少低载畜能力地区的压力。③ 这样的季节性迁徙行为体现了一种适应当地自然资源条件的传统牧业经营管理模式。

对于无法承担昂贵气候变化适应技术的牧民和农牧民而言,有效且易负担的适应措施包括:其一,为牲畜提供遮阳和饮用水以减轻温度升高造成的热应激反应。考虑到目前较高的能源价格,为牲畜提供自然(低成本)的遮阳以替代高成本的空调制冷设施,可以有效降低饲养成本。其二,缩小畜群规模。数量较少但高产的畜群能够提高畜牧生产效率且减少其所产生的温室气体排放。其三,改变畜群结构,选择大型牲畜而非小型牲畜。其四,改善水源管理,推广简单有效的灌溉技术(如滴灌与喷灌)以及雨水搜集存储设施。

(10) 用多年生牧草代替一年生牧草 (conversion from annual to perennial pastures)。用苜蓿等多年生牧草代替一年生牧草可以使土壤得到有效改良,土壤中固氮量得到增加,减少农地氮肥的使用;同时,多年生牧草的水土保持能力要高于一年生作物,向多年生牧草的转化可以有效降低土壤侵蚀,减

---

① 李胜利、金鑫、范学珊等:《反刍动物生产与碳减排措施》,载《动物营养学报》2010 年第 1 期。

② Hoffmann, I. Livestock genetic diversity and climate change adaptation [A]. In: Rowlinson, P., Steele, M., Nefzaoui, A. (Eds.), Proceedings of the International Conference Livestock and Global Climate Change [C]. Hammamet, Tunisia, 2008.

③ Akinnagbe, O. M., Irohibe, I. J. Agricultural adaptation strategies to climate change impacts in Africa: a review [J]. Bangladesh Journal of Agricultural Research, 39 (3): 407–418.

少氮的流失量，保持土壤墒情。①

以上应对措施既可单独进行也可运用其中几项进行组合搭配。

绝大多数现有文献采用由实证研究所获得的参数来对农业温室气体排放量进行估计。目前学术界所使用的绝大部分排放参数是在北半球农业环境中获得的。② 对于南半球农业环境下的排放参数估计较少。

很多文献都采用由联合国政府间气候变化专门委员会所提供的评价方法（简称 IPCC 估计法）来估计农业温室气体的排放量。IPCC 估计法为不同的温室气体定义了具体减排措施的排放因子（emissions factors），然后基于排放量与排放源之间的线性关系计算具体的温室气体排放量。将不同排放源所产生的温室气体排放量进行加总，可以得到农业系统的温室气体总排放量。③

在具体应用方面，杜伽哈（De Cara）以及杜伽哈和热耶（De Cara and Jayet et al.）利用 IPCC2001 估计法，分析了欧盟地区旨在减少农业甲烷与氧化亚氮排放的农业减排措施。他们总共考虑了 11 种农业温室气体排放源。④ 贝卡姆等（Bakam et al.）和波涅斯莫等（Bonesmo et al.）利用 IPCC2006 估计法，分别分析了苏格兰的混合农业与挪威种植农业的温室气体减排活动。⑤

另外一个被广泛运用的途径是利用 CENTURY 模型来模拟计算农业温室气体的排放量。CENTURY 是一个一般化的、考虑作物生态系统的生物地球

---

① 杨景成、韩兴国、黄建辉等：《土壤有机质对农田管理措施的动态响应》，载《生态学报》2003 年第 4 期。Kragt, M. E., Pannell, D. J., Robertson, M. J., Thamo, T. Assessing costs of soil carbon sequestration by crop-livestock farmers in Western Australia [J]. Agricultural Systems, 2012, 112: 27 – 37. Thamo, T., Addai, D., Pannell, D. J., et al. Climate change impacts and farm-level adaptation: Economic analysis of a mixed cropping-livestock system [J]. Agricultural Systems, 2017, 150: 99 – 108.

② Barton, L., Kiese, R., Gatter, D., et al. Nitrous oxide emissions from a cropped soil in a semi-arid climate [J]. Global Change Biology, 2008, 14: 177 – 192.

③ IPCC. 2006 Guidelines for National Greenhouse Gas Inventories. Intergovernmental Panel on Climate Change [R]. National Greenhouse Gas Inventories Programme, IPCC, 2006.

④ De Cara, S., Houzé, M., Jayet, P. A. Methane and nitrous oxide emissions from agriculture in the EU: A spatial assessment of sources and abatement costs [J]. Environmental and Resource Economics, 2005, 32 (4): 551 – 583. De Cara, S., Jayet, P. A. Marginal abatement costs of greenhouse gas emissions from European agriculture, cost effectiveness, and the EU non-ETS burden sharing agreement [J]. Ecological Economics, 2011, 70 (9): 1680 – 1690.

⑤ Bakam, I., Balana, B. B., Matthews, R. Cost-effectiveness analysis of policy instruments for greenhouse gas emission mitigation in the agricultural sector [J]. Journal of Environmental Management, 2012, 112: 33 – 44. Bonesmo, H., Skjelvåg, A. O., Henry Janzen, H., et al. Greenhouse gas emission intensities and economic efficiency in crop production: A systems analysis of 95 farms [J]. Agricultural Systems, 2012, 110: 142 – 151.

化学模型。① 该模型包括了土壤生物地球化学、作物生长与产量、植被（草地、森林与热带草原）和水热平衡等子模块，可用于研究包括轮作、肥料管理、耕作法、肥料管理和放牧管理等农业减排措施的碳、氮以及其他营养物的动态过程。

一系列的研究利用 CENTURY 模型分析了美国中部地区的一个旱地谷类生产系统。这些研究分别考虑了包括持续性耕作、退耕还草、保护性耕作与减少休耕频率在内的碳汇农业实践。② 贡萨雷斯埃斯特拉达等（González-Estrada et al.）整合运用 CENTURY 和一个旨在分析农业科技转移的决策支持系统模块，研究了持续性耕作、轮作与减少休耕频率等减排措施在加纳热带草原农业系统的应用。③

2.2 气候变化所引起的制度环境方面的改变

为了应对气候变化，世界各国政府纷纷设定了多种的温室气体减排目标。例如，美国政府在联合国气候变化框架公约中保证其 2020 年排放量比 2005 年的水平下降 17%。④ 2009 年，中国政府提出，到 2020 年单位国内生产总值

---

① Antle, J. M., Capalbo, S. M., Mooney, S., et al. Economic analysis of agricultural soil carbon sequestration: an integrated assessment approach [J]. Journal of Agricultural and Resource Economics, 2001, 26 (2): 344 – 367.

② Antle, J. M., Capalbo, S. M., Mooney, S., et al. Sensitivity of carbon sequestration costs to soil carbon rates [J]. Environmental Pollution, 2002, 116 (3): 413 – 422. Antle, J. M., Capalbo, S. M., Mooney, S., et al. Spatial heterogeneity, contract design, and the efficiency of carbon sequestration policies for agriculture [J]. Journal of Environmental Economics and Management, 2003, 46 (2): 231 – 250. Capalbo, S. M., Antle, J. M., Mooney, S., et al. Sensitivity of carbon sequestration costs to economic and biological uncertainties [J]. Environmental Management, 2004, 33 (1): S238 – S251. Mooney, S., Antle, J., Capalbo, S., et al. Influence of project scale and carbon variability on the costs of measuring soil carbon credits [J]. Environmental Management, 2004, 33 (1): S252 – S263. Antle, J. M., Capalbo, S. M., Paustian, K., et al. Estimating the economic potential for agricultural soil carbon sequestration in the Central United States using an aggregate econometric-process simulation model [J]. Climatic Change, 2007, 80 (1 – 2): 145 – 171.

③ González-Estrada, E., Rodriguez, L. C., Walen, V. K., et al. Carbon sequestration and farm income in West Africa: Identifying best management practices for smallholder agricultural systems in northern Ghana [J]. Ecological Economics, 2008, 67 (3): 492 – 502.

④ https://www.whitehouse.gov/the-press-office/2015/03/31/fact-sheet-us-reports-its-2025-emissions-target-unfccc. 需要说明的是，美国政府已于 2019 年 11 月 4 日正式宣布退出应对全球气候变化的《巴黎协定》，预计将于 2020 年 11 月 4 日生效。参见 https://news.un.org/zh/story/2019/11/1044981。

二氧化碳排放比2005年降低40%～45%。① 在澳大利亚，联邦政府在制定了到2020年减少5%的温室气体排放目标的基础上，承诺2050年前将澳大利亚的总排放量从2000年的水平减少60%。② 为了实现减排目标，各国正不断地构想与实施温室气体减排措施，以应对气候变化。

目前各国所采用的碳定价机制大多涵盖了电力生产部门以及主要的能源密集型产业，并没有将农业纳入其中。在这些机制中，学界认为碳税、总量管制与排放交易和碳减排信用可以运用于农业领域温室应对气候变化的实践。

碳税被认为是碳定价机制中最简单的措施，③ 目前已被广泛应用于一些欧盟国家（如瑞典、芬兰、荷兰和丹麦）以及加拿大的不列颠哥伦比亚省。在一个碳税体系下，市场主体需要对其排放的每吨温室气体缴税。在理性经济人的前提下，减排成本低的市场主体会倾向于通过减少排放而支付更少的税，而减排成本高的市场主体会选择继续排放温室气体并为其排放缴税。最终碳税有效地成为最终产品价格的一个组成部分。

总量管制与排放交易亦被用作一种减排政策工具。在总量管制与排放交易体系中，总体排放数量是被限定的；排放者在限定的总量下被分配若干排放许可或配额，并且能够对其进行交易。澳大利亚政府在2012～2015年实施的碳价计划（carbon pricing scheme）就是一项典型的总量管制与排放交易措施。该计划规定澳大利亚所有温室气体年排放量超过2.5万 tCO_2e 的企业均需购买排放许可。单位碳减排信用是指"对于在所要求的排放控制水平基础上自愿减少的排放量依要求给予的一个额度"④。

碳减排信用可以被租赁、交易或储蓄以供未来使用。⑤ 如果总量管制与排放交易被归为一种自上而下的减排政策，那么碳减排信用则恰恰相反，是一种自下而上的政策。换言之，碳减排信用体系能鼓励未受监管的主体或公司为那些在监管下超额排放的企业提供排放信用。⑥ 这些碳排放信用体系经常被国内总量管制与排放交易系统所涵盖。澳大利亚政府目前正在规划的减

---

① http://finance.people.com.cn/GB/10461522.html.
② Garnaut, R. The Garnaut Climate Change Review [M]. Cambridge: Cambridge University Press, 2008.
③ Metcalf, G. E. A proposal for a US carbon tax swap: An equitable tax reform to address global climate change [R]. Discussion Paper 2007 – 12, Hamilton Project, Brookings Institute, 2007.
④⑤ http://www.swcleanair.org/erc.html.
⑥ 贾敬敦、魏珣、金书秦：《澳大利亚发展碳汇农业对中国的启示》，载《中国农业科技导报》2012年第2期。

少排放量基金就属于碳减排信用政策。

按照总量管制与排放交易和碳减排信用这两种方式，温室气体的排放成为一种可供买卖的商品，市场机制决定了温室气体排放的价格与交易量。在此种情况下，排放额度及其市场价格完全由市场对其的供需关系所决定，可高可低。由于交易价格是由市场机制所决定的，整个交易过程可以保障排放者以较低的成本高效地完成减排或排放转移，同时也减少了整个社会的减排及气候变化成本。

澳大利亚联邦政府于 2011 年 7 月 10 日正式公布了一项碳价计划，并自 2012 年 7 月 1 日起在全澳范围内正式实施。该计划由两阶段所构成：计划的前三年是固定价格阶段，接下来是浮动价格阶段。在固定价格阶段，温室气体的价格为 2012~2013 财年 23 澳元/tCO_2e，2013~2014 财年 24.15 澳元/tCO_2e，2014~2015 财年 25.4 澳元/tCO_2e。在浮动价格阶段（开始于 2015 年 7 月 1 日），温室气体的价格将由市场决定。大部分的排放信用将由澳大利亚的碳价计划管理机构——清洁能源监管局（Clean Energy Regulator）以拍卖的方式进行分配。每年度全澳的排放配额总量将由政府设定，并受到由监管机构设定的排放总量上限的制约。在该政策执行期间，全澳年排放量超过限额（2.5 万 tCO_2e）的企业，将为其排放的温室气体购买排放配额，或支付差额费用。

虽然雨养农业尚未被该计划所涵盖，但澳大利亚一直存在扩大碳价计划适用范围的声音。倘若雨养农业被纳入该计划中，雨养农业经营者将不得不为其过量排放的温室气体支付费用。此时减少温室气体排放量对于提高农业生产利润、保证雨养农业的可持续发展的意义将更加突出。此外，对于澳大利亚的雨养农民而言，该计划的实施使得其通过出售由农业温室气体减排与固碳所赚取的排放信用来赚取额外收入成为可能。此外，作为在碳价计划交易市场中排放信用的潜在来源，雨养农业减排与固碳活动有可能降低实现减排目标的社会成本。[①]

自澳大利亚自由党艾伯特政府上台后，为了兑现选举承诺，削减相关行业生产成本，澳大利亚联邦政府开始寻求对于碳价计划的替代项目。2013 年 11 月 21 日，澳大利亚联邦众议院投票通过碳税废除法案；2014 年 7 月 17

---

① Bernoux, M., Cerri, C. C., Volkoff, B., et al. Gaz à effet de serre et stockage du carbone par les sols, inventaire au niveau du Brésil [J]. Cahiers Agriculrures, 2005, 14: 96-100.

日，澳大利亚联邦参议院以39票赞成、32票反对，通过废除碳税的系列法案。目前澳大利亚联邦政府正在寻求以类似于碳交易系统的方式替代碳价计划。

2013年12月20日，澳大利亚联邦政府正式公布了《减少排放量基金绿皮书》（Emissions Reduction Fund Green Paper），宣布澳大利亚计划建立一个全国性的温室气体减排基金，为低成本的减排活动提供资金支持。① 该基金计划依照确定的基准价（benchmark price）对碳减排信用进行拍卖。此基准价是支付给每吨减少的温室气体款项的上限。只有减排成本低于此基准价的减排计划才有资格进入拍卖。该基金正式建立以后，农业经营者减排活动的成本如具有比较优势，将有机会直接从政府获得资金扶持。这为农民增收又创造了一条潜在的途径。

为了更好地鼓励农业减排，澳大利亚联邦政府于2011年开始陆续出台了《碳信用额（碳汇农业方案）法案2011》[Carbon Credits (Carbon Farming Initiative) Act 2011，简称碳汇农业方案] 及其修正案②，明确了农业减排补偿项目的适用对象、途径、信息公开和审计等内容。该方案为农民提供了额外的增收机会：农业经营者可以通过改变农业土地利用及生产方式来减少温室气体排放与固碳，从而赚取排放信用；随之可将其所赚取的排放信用出售给那些需要购买排放信用的企业，以赚取额外收入。③

2011年，中国国家发展改革委员会宣布在五个城市（北京、上海、天津、重庆、深圳）和两个省（湖北、广东）实施七项区域碳排放交易试点项目。2013~2014年，七项区域碳排放交易试点项目陆续启动运行。这七个试点项目覆盖2.6亿人口，能源消耗量为8.3亿tCO_2e，14万亿元人民币GDP。目前这些项目都只针对二氧化碳排放。中国各地试点计划的总碳排放额度仅次于欧盟碳排放交易机制。中国碳排放交易机制也很可能成为规模日渐增长的亚太区碳交易市场中最重要的中心。④ 中国的碳排放交易项目，是目前全球在发展中经济体中唯一实际运行的碳交易市场。

---

① Australian Government. Emissions Reduction Fund Green Paper [R]. Commonwealth of Australia, 2014.

② 最新修订的《碳信用额（碳汇农业方案）法案2011》（编号：C2016C00029）于2015年12月10日开始实施。

③ Clean Energy Regulator. Carbon pricing mechanism [R]. Australian Clean Energy Regulator, 2012.

④ https：//www.lw.com/admin/Upload/Documents/China-Carbon-Trading_Chinese.pdf.

原则上，国家发展改革委员会负责监管所有试点项目的设计及实施。事实上，国家发展改革委员会将监管权力下放到省级和市级的发展改革委员会来制定自己的制度。因此，不同的试点项目会有不同的实施计划、法规及政府规范。不同试点项目的关键特点，包括配额设置、覆盖产业、配额分配、计算标准、检测、报告和核实规则以及可交易产品都有所不同。例如，在不同的试点计划下，有的省份纯粹用排放量决定规管范围（如北京仅将二氧化碳年排放量超过1万吨的企业纳入计划），有的省份则会按照行业板块等附加参照制定排放水平（上海为石化及航空航天行业设定不同的限额水平）。试点的一个突出特点就是计算中包含两种类型的排放，像发电站排放类别的直接排放和间接排放，如以热量和电力消费为表现形式的排放。这种计算方法会导致某些排放源出现重复计算的情况，但是在现阶段，这样的计算方式是必要的，因为中国的电价是严格管控的。① 如果近期能对电力行业进行市场改革，那么该国家项目将无须覆盖间接排放。

这七项试点项目自建立以来均实现了多个政策目标。覆盖了试点地区的大部分排放产业和企业，在实现地方碳强度目标和控制温室气体排放方面发挥了重要作用。项目的运营帮助各地区建立了技术基础和制度能力经验，还塑造了首个碳排放市场，也形成了中国碳排放价格。通过这些项目，企业在减排和提高公众对气候变化意识方面的作用得到大幅度提高。尽管如此，试点项目还是存在许多突出问题。不同试点项目在登记和交易标准、计算、MRV规则及其他方面的差距给未来国家项目的连接性和实施程度带来了挑战。②

2015年9月25日，习近平主席在与美国时任总统奥巴马发布的两国气候变化声明中宣布，中国将于2017年启动全国碳交易体系。有别于试点计划，全国碳交易机制草案的范围覆盖七种温室气体：二氧化碳（CO_2）、甲烷（CH_4）、氧化亚氮（N_2O）、氢氟烃（HFCs）、全氟化合物（PFCs）、六氟化硫（SF_6）及三氟化氮（NF_3）。计划下的交易产品有两种：碳排放额及核定自愿减排量。

碳排放额是分配给受碳交易机制碳排放上限规范的企业，机制下的企业或其他市场参与者可以买卖碳排放额。企业每持有一单位的碳排放额可排放

---

①② https：//www.belfercenter.org/sites/default/files/files/publication/CN% 20HKS - Tsinghua% 20Workshop% 202016.pdf.

1tCO$_2$e 温室气体。国家发展改革委员会负责决定机制下的总碳排放额,并根据各辖区的温室气体排放历史数据、经济增长,以及产业和能源结构决定分配给各省区市的碳排放额。各省级政府再以免费和收费的方式向辖区内所涵盖企业分配碳排放额,其中,收费碳排放额将扮演日益重要的角色。中国碳排放交易机制初期将涉及发电、钢铁、化工、建材、水泥、造纸和有色金属制造行业。这些行业纳入近 40 亿 tCO$_2$e 的温室气体排放,相当于全国总排放量的近 40%。

核定自愿减排量可用于抵消超出上限的每单位 tCO$_2$e,日后也有可能用于中国境外,等同于北美的抵扣额或者欧盟碳排放交易机制和清洁发展机制下的核定减排量。在中国碳交易机制下,企业可用核定自愿减排量抵消排放量,这可以避免出现价格在庞大需求下急升的情况,也可以为减排面临较大挑战的行业企业提供弹性。企业和个人都将被允许参与碳交易,这样的公开市场将有助于形成合理的价格。

虽然中国雨养农业尚未被碳交易计划或碳税所涵盖,但考虑到雨养农业每年排放了大量温室气体,未来存在将雨养农业纳入碳税或碳交易计划的可能。[①] 倘若雨养农业被纳入碳税中,雨养农业经营者将不得不为其过量排放的温室气体支付费用。此时,减少温室气体排放量对于提高雨养农业生产利润、有效应对气候变化、保证雨养农业可持续发展的意义将更加突出。此外,对于中国的雨养农业经营者而言,全国碳交易市场的实施使得其通过出售由农业温室气体减排与固碳所赚取的排放信用来赚取额外收入成为可能。作为在中国碳交易市场中排放信用的潜在来源,雨养农业减排与固碳活动有可能降低实现中国减排目标、实现对气候变化有效应对的社会成本。

在实证研究中,政策背景是研究气候变化条件下雨养农业的重要组成部分。目前,现实环境中绝大多数国家市场型减排政策都没有涵盖农业领域为应对气候变化所采取的温室气体减排活动。新西兰是唯一的例外。[②] 作为新西兰的第一大温室气体排放来源,农业产生了占该国总排放量 55% 的温室气体。[③]

---

① Tang, K., He, C., Ma, C., Wang, D. Does carbon farming provide a cost-effective option to mitigate GHG emissions? Evidence from China [J]. Australian Journal of Agricultural and Resource Economics, 2019, 63 (3): 575-592. Tang, K., Hailu, A. Smallholder farms' adaptation to the impacts of climate change: Evidence from China's Loess Plateau [J]. Land Use Policy, 2020, 91, 104353.

② 在撰写本书时,笔者尚未发现有其他国家规定农业需要为其排放的温室气体付费。

③ O'Hara, P., Freney, J., Ulyatt, M. Abatement of agricultural non-carbon dioxide greenhouse gas emissions [R]. Report for the Ministry of Agriculture and Forestry, 2003.

鉴于此，该国已经通过立法规定自2015年起农业将被强制纳入一个基于市场的减排系统中。① 鉴于此，有关的经济分析大多考虑潜在的政策情景。这些政策情景可分为三种，即按量合约（碳信用）、按面积合约和基于成本的合约。

许多研究考虑了一种基于市场、产出导向的按量合约。此种合约假设，农民通过改变土地利用及农场管理活动来减少温室气体排放，并依据减排的数量获得碳信用；然后政府或碳市场按照碳信用的数量支付给农民一定金额的款项。此处碳信用被假定为可以交易的商品。已有文献多假设碳信用的价格在一定的区间内波动；每次波动的数量可以是一个固定值，或者是一个变化值。例如，安托等（Antle et al.）假设每 tCO_2e 的价格在 10~100 美元变动，每次的变化量为 10 美元。② 杜伽哈和热耶（De Cara and Jayet）描述了一种每 tCO_2e 价格在 0~10000 欧元波动的情景，且单次的变化量不断增加。③ 许多文献都采取了类似的政策情景假设。④

一些研究分析了一种投入导向的按面积合约。在此种政策情景下，农民依照被纳入减排计划的土地面积数从政府领取补偿款。安托等（Antle et al.）设计了一种按每年每公顷土地支付固定金额的政策情景。该情景中，补偿款项在每公顷 5~50 美元不等，按每次 5 美元的幅度增加。⑤ 有许多研究都采纳了这一情景。补偿通常以现金形式支付，但有时也以现金加实物的方式提供

---

① Cooper, M. H., Boston, J., Bright, J. Policy challenges for livestock emissions abatement: lessons from New Zealand [J]. Climate Policy, 2013, 13 (1): 110-133.

② Antle, J. M., Capalbo, S. M., Mooney, S., et al. Spatial heterogeneity, contract design, and the efficiency of carbon sequestration policies for agriculture [J]. Journal of Environmental Economics and Management, 2003, 46 (2): 231-250.

③ De Cara, S., Jayet, P. A. Marginal abatement costs of greenhouse gas emissions from European agriculture, cost effectiveness, and the EU non-ETS burden sharing agreement [J]. Ecological Economics, 2011, 70 (9): 1680-1690.

④ Pendell, D. L., Williams, J. R., Boyles, S. B., et al. Soil carbon sequestration strategies with alternative tillage and nitrogen sources under risk [J]. Applied Economic Perspectives and Policy, 2007, 29 (2): 247-268. Grace, P. R., Antle, J., Ogle, S., et al. Soil carbon sequestration rates and associated economic costs for farming systems of south-eastern Australia [J]. Australian Journal of Soil Research, 2010, 48: 1-10. Kragt, M. E., Pannell, D. J., Robertson, M. J., et al. Assessing costs of soil carbon sequestration by crop-livestock farmers in Western Australia [J]. Agricultural Systems, 2012, 112: 27-37.

⑤ Antle, J. M., Capalbo, S. M., Mooney, S., et al. Economic analysis of agricultural soil carbon sequestration: an integrated assessment approach [J]. Journal of Agricultural and Resource Economics, 2001, 26 (2): 344-367.

给农户。① 还有学者提到了中国的退耕还林计划。中央政府每年以每亩土地补偿150千克的粮食和20元现金的标准补偿5~8年。②

有些研究也探讨了基于成本、投入导向的合约。这些合约假设依照农业减排活动的机会成本与转化成本为农民提供补偿。帕克斯和哈迪（Parks and Hardie）设计了一种退耕还林的合约。该合约依照按植树成本的一半与土地年均出租价格之和为农民提供10年的补偿。③ 泽勒克和希乌力（Zelek and Shively）构想了一种基于机会成本与农林地转化成本之和的连续的补偿情景。④

绝大部分的实证研究仅考虑了单一的政策情景。但也有一些研究涵盖了两种不同的政策情景，并对它们进行了成本效益分析。安托等（Antle et al.）比较了一种固定按面积合约与一种按量合约的效益。他们发现，两种合约的效率以及减排测量成本都与空间异质性正向相关；在考虑测量成本之后，按量合约的效益似乎会比按面积合约的效益更高。⑤ 贝卡姆等（Bakam et al.）模拟了一种按量合约和一种碳信用交易系统。他们认为，政策情景的成本有效性取决于激励的水平以及分配给农民的免费碳信用的数量。⑥

2.3　气候变化对于雨养农业的影响

近年来，气候变化日益成为当今全球研究的热点问题，其影响包括了环

---

① Antle, J. M., Capalbo, S. M., Mooney, S., et al. Sensitivity of carbon sequestration costs to soil carbon rates [J]. Environmental Pollution, 2002, 116 (3): 413 – 422. Antle, J. M., Capalbo, S. M., Mooney, S., et al. Spatial heterogeneity, contract design, and the efficiency of carbon sequestration policies for agriculture [J]. Journal of Environmental Economics and Management, 2003, 46 (2): 231 – 250.

② Xu, W., Yin, Y., Zhou, S. Social and economic impacts of carbon sequestration and land use change on peasant households in rural China: A case study of Liping, Guizhou Province [J]. Journal of Environmental Management, 2007, 85 (3): 736 – 745.

③ Parks, P. J., Hardie, I. W. Least-cost forest carbon reserves: Cost-effective subsidies to convert marginal agricultural land to forests [J]. Land economics, 1995, 71 (1): 122 – 136.

④ Zelek, C. A., Shively, G. E. Measuring the opportunity cost of carbon sequestration in tropical agriculture [J]. Land Economics, 2003, 79 (3): 342 – 354.

⑤ Antle, J. M., Capalbo, S. M., Mooney, S., et al. Spatial heterogeneity, contract design, and the efficiency of carbon sequestration policies for agriculture [J]. Journal of Environmental Economics and Management, 2003, 46 (2): 231 – 250.

⑥ Bakam, I., Balana, B. B., Matthews, R. Cost-effectiveness analysis of policy instruments for greenhouse gas emission mitigation in the agricultural sector [J]. Journal of Environmental Management, 2012, 112: 33 – 44.

境、生态、经济、社会、文化、健康、能源和农业生产等诸多方面。① 雨养农业作为对气候变化最为敏感的产业,是受气候变化影响最深远、最直接的生产行业之一。光、热、水、气等气候因素的变化,直接引起雨养农业生产条件和产出水平的变动。②

大多数学者的研究区域都集中在欧美澳等发达国家和地区,他们从作物产量③、农业种植结构④、农业用地分配⑤、农户利润⑥、农业温室气体排

---

① 初征、郭建平、赵俊芳:《东北地区未来气候变化对农业气候资源的影响》,载《地理学报》2017年第7期。IPCC. Global Warming of 1.5℃ [R]. IPCC, 2018. Tang, K., Hailu, A., Kragt, M. E., Ma, C. The response of broadacre mixed crop-livestock farmers to agricultural greenhouse gas abatement incentives [J]. Agricultural Systems, 2018, 160: 11 – 20.

② 胡慧芝、刘晓琼、王建力:《气候变化下汉中盆地水稻产量变化研究》,载《自然资源学报》2018年第4期。刘玉洁、陈巧敏、葛全胜等:《气候变化背景下1981~2010中国小麦物候变化时空分异》,载《中国科学:地球科学》2018年第7期。Tang, K., Hailu, A., Kragt, M. E., Ma, C. The response of broadacre mixed crop-livestock farmers to agricultural greenhouse gas abatement incentives [J]. Agricultural Systems, 2018, 160: 11 – 20. Tang, K., Hailu, A. Smallholder farms' adaptation to the impacts of climate change: Evidence from China's Loess Plateau [J]. Land Use Policy, 2020, 91: 104353.

③ Thamo, T., Addai, D., Pannell, D. J., et al. Climate change impacts and farm-level adaptation: Economic analysis of a mixed cropping-livestock system [J]. Agricultural Systems, 2017, 150: 99 – 108. Bonan, G. B., Doney, S. C. Climate, ecosystems, and planetary futures: The challenge to predict life in Earth system models [J]. Science, 2018, 359 (6375): eaam8328. Tang, K., Hailu, A., Kragt, M. E., Ma, C. The response of broadacre mixed crop-livestock farmers to agricultural greenhouse gas abatement incentives [J]. Agricultural Systems, 2018, 160: 11 – 20.

④ Kragt, M. E., Pannell, D. J., Robertson, M. J., Thamo, T. Assessing costs of soil carbon sequestration by crop-livestock farmers in Western Australia [J]. Agricultural Systems, 2012, 112: 27 – 37. Powell, J. W., Welsh, J. M., Eckard, R. J. An irrigated cotton farm emissions case study in NSW, Australia [J]. Agricultural Systems, 2017, 158: 61 – 67.

⑤ Thamo, T., Addai, D., Pannell, D. J., et al. Climate change impacts and farm-level adaptation: Economic analysis of a mixed cropping-livestock system [J]. Agricultural Systems, 2017, 150: 99 – 108. Antle, J. M., Zhang, H., Mu, J. E., et al. Methods to assess between-system adaptations to climate change: dryland wheat systems in the Pacific Northwest United States [J]. Agriculture, Ecosystems & Environment, 2018, 253: 195 – 207. Antle, J. M., Cho, S., Tabatabaie, S. M. H., et al. Economic and environmental performance of dryland wheat-based farming systems in a 1.5℃ world [J]. Mitigation and Adaptation Strategies for Global Change, 2019, 24 (2): 165 – 180.

⑥ Kragt, M. E., Pannell, D. J., Robertson, M. J., Thamo, T. Assessing costs of soil carbon sequestration by crop-livestock farmers in Western Australia [J]. Agricultural Systems, 2012, 112: 27 – 37. Antle, J. M., Zhang, H., Mu, J. E., et al. Methods to assess between-system adaptations to climate change: dryland wheat systems in the Pacific Northwest United States [J]. Agriculture, Ecosystems & Environment, 2018, 253: 195 – 207.

放量①等方面探讨了气候变化对于雨养农业的影响。总体而言，这些研究认为，当前气候变化给雨养农业带来的负面影响和威胁远大于正面影响，雨养农业的诸多方面在气候变化条件下都变得更加脆弱。

近年来，一些学者逐渐开始关注气候变化对发展中国家农业的影响。② 现有研究认为，发展中国家雨养农业多处于局部地形与气候多变地区，不同季节水蚀风蚀交替出现，存在水土流失现象，生态环境脆弱，农业生产力水平偏低，极易受气候变化的影响。③ 此外，这些区域多为经济社会欠发达地区，当地农业基础设施薄弱，农户缺乏生产性资金，农业生产技术落后，贫困和疾病发生率偏高。④ 尽管许多发展中国家在提高农业生产力和改善环境条件方面取得了长足的进步，撒哈拉以南的非洲和亚洲许多国家众多从事雨养农业生产活动的农户依然面临着贫困、饥饿、食物紧缺以及营养不良问题。⑤ 全球超过一半的饥饿人口以及超过70%的营养不良儿童都生活在主要从事雨养

---

① Tang, K., Hailu, A., Kragt, M. E., Ma, C. Marginal abatement costs of greenhouse gas emissions: broadacre farming in the Great Southern Region of Western Australia [J]. Australian Journal of Agricultural and Resource Economics, 2016, 60 (3): 459 – 475. Meier, E. A., Thorburn, P. J., Kragt, M. E., et al. Greenhouse gas abatement on southern Australian grains farms: Biophysical potential and financial impacts [J]. Agricultural Systems, 2017, 155: 147 –157.

② 王莉：《雨养农业对气候变化脆弱性的实证研究——以甘肃省华池县为例》，载《农村经济》2013年第3期。Khanal, U., Wilson, C., Hoang, V. N., et al. Farmers' adaptation to climate change, its determinants and impacts on rice yield in Nepal [J]. Ecological Economics, 2018, 144: 139 –147. Agriculture, Ecosystems & Environment, 2016, 220: 164 – 174. Powlson, D. S., Stirling, C. M., Thierfelder, C., et al. Does conservation agriculture deliver climate change mitigation through soil carbon sequestration in tropical agro-ecosystems? [J]. Agriculture, Ecosystems & Environment, 2016, 220: 164 – 174. 姚玉璧、杨金虎、肖国举等：《气候变暖对西北雨养农业及农业生态影响研究进展》，载《生态学杂志》2018年第7期。马雅丽、郭建平、赵俊芳：《晋北农牧交错带作物气候生产潜力分布特征及其对气候变化的响应》，载《生态学杂志》2019年第3期。

③ Eldoma, I. M., Li, M., Zhang, F., Li, F. M. Alternate or equal ridge-furrow pattern: Which is better for maize production in the rain-fed semi-arid Loess Plateau of China? [J]. Field Crops Research, 2016, 191: 131 –138. 田贵良、贾琨颢、孙兴波等：《干旱事件影响下虚拟水期权契约的提出及其定价研究》，载《农业技术经济》2016年第9期。石晓丽、史文娇：《北方农牧交错带界线的变迁及其驱动力研究进展》，载《农业工程学报》2018年第20期。

④ 谭淑豪、谭文列婧、励汀郁等：《气候变化压力下牧民的社会脆弱性分析——基于内蒙古锡林郭勒盟4个牧业旗的调查》，载《中国农村经济》2016年第7期。王一超、郝海广、张惠远等：《农牧交错区农户生计分化及其对耕地利用的影响——以宁夏盐池县为例》，载《自然资源学报》2018年第2期。

⑤ Rockström, J., Falkenmark, M. Agriculture: increase water harvesting in Africa [J]. Nature, 2015, 519 (7543): 283 –285.

农业生产活动的地区。① 总体而言，这些研究的研究对象多集中在对于农业气象条件、单一作物产量和作物生长状态变量的作用，而对雨养农业在气候变化所引起环境与社会方面影响综合作用下的种—畜结构、粮食作物与经济作物比例、农地分配、农户收益、农业温室气体排放等问题鲜有涉及。

已有研究多关注气候变化对种植业的影响。现有研究表明，受全球表面平均温度上升的影响，一些区域的作物产量出现了下降，并且这一趋势将随着平均温度的进一步上升而持续。② 气候变化所引起的农业病虫害会造成小麦、水稻以及玉米等作物的减产，全球表面平均温度每上升1℃会使得农业病虫害造成的作物产量损失增加10%～25%。③ 在美国雨养农业区，高温天气的增加导致平均玉米生产潜力下降了35%。④ 在非洲雨养农业区，农业产量受气候变化和变异的严重影响，尤其是干旱和半干旱地区的宜农土地、生长季和潜在产量均有所下降。截至2020年，一些国家的雨养农业产量将下降最高达50%。⑤ 全球平均气温每升高1℃，非洲干旱地区的农业收成将减少约10%。⑥ 非洲玉米产量对热量响应的主要影响因素是水分，即在雨量充沛的条件下，日气温超过30℃时，气温每增加1℃产量会降低1%；在干旱条件下，气温每增加1℃产量降低1.7%。⑦

在中国雨养农业区，平均温度的上升使得小麦和玉米的产量偏低10%以上。⑧

---

① FAO, IFAD, UNICEF, WFP, WHO. The state of food security and nutrition in the world 2018: building climate resilience for food security and nutrition [R]. FAO, Rome, 2018.

② IPCC. IPCC Fifth Assessment Report (AR5) [M]. Cambridge: Cambridge University Press, 2014. 戴彤、王靖、赫迪等：《基于APSIM模型的气候变化对西南春玉米产量影响研究》，载《资源科学》2016年第1期。Fang, J., Yu, G., Liu, L., Hu, S., Chapin Ⅲ, F. S. Climate change, human impacts, and carbon sequestration in China [J]. Proceedings of the National Academy of Sciences, 2018, 115 (16): 4015-4020.

③ Deutsch, C. A., Tewksbury, J. J., Tigchelaar, M., et al. Increase in crop losses to insect pests in a warming climate [J]. Science, 2018, 361 (6405): 916-919.

④ Huffman, W. E., Jin, Y., Xu, Z. The economic impacts of technology and climate change: new evidence from US corn yields [J]. Agricultural Economics, 2018, 49 (4): 463-479.

⑤⑥ 史文娇、陶福禄：《非洲农业产量对气候变化响应与适应研究进展》，载《中国农业科学》2014年第16期。

⑦ Lobell, D. B., Bänziger, M., Magorokosho, C., Vivek, B. Nonlinear heat effects on African maize as evidenced by historical yield trials [J]. Nature Climate Change, 2011, 1 (1): 42-45.

⑧ 马玉平、孙琳丽、俄有浩等：《预测未来40年气候变化对我国玉米产量的影响》，载《应用生态学报》2015年第1期。陈帅、徐晋涛、张海鹏：《气候变化对中国粮食生产的影响——基于县级面板数据的实证分析》，载《中国农村经济》2016年第5期。初征、郭建平、赵俊芳：《东北地区未来气候变化对农业气候资源的影响》，载《地理学报》2017年第7期。

杨轩结合大田试验数据和作物模型（APSIM），依据不同气候变化情景，对黄土高原雨养农业长期尺度下传统作物系统和改进的作物系统的生产进行了模拟。结果显示，在气温升高、降水量减少的情景下，冬小麦、玉米和紫花苜蓿产量降低，最大幅度分别可达38.7%、40.3%和41.8%。冬小麦和紫花苜蓿在气温降低、降水量增加时增产，最大增幅分别为29.8%和51.7%。玉米在降水量增加、温度不变的情景下增产幅度最大，为22.0%。在设定范围内，紫花苜蓿的产量变异范围受气候变化的影响最小，说明其适应能力较强。① 何亮分析了黄土高原气象数据、农业气象试验站观测数据、以及长尺度产量、水分管理等数据，发现黄土高原区域的冬小麦播种、出苗、越冬开始分别平均推迟了1.2、1.3、1.2天/10年；返青、开花、成熟期平均分别提前了2.0、3.7、3.1天/10年；生育期、越冬期（越冬开始到返青开始）和出苗到开花（营养生长）平均分别缩短了4.3、3.1和5.0天/10年，而开花到成熟（生殖生长）延长了0.7天/10年。②

目前学界对气候变化条件下雨养农业区畜牧业的关注较少。雨养农业由于多分布于干旱半干旱地区，农业生产普遍采用同时种植作物和从事畜牧养殖的混合经营模式。现有研究表明，畜牧业是最大的非碳温室气体排放来源，全球每年平均排放31亿 tCO_2e 的甲烷和20亿 tCO_2e 的氧化亚氮，分别占全球总排放量的44%和53%③。气候变化条件下的气温升高、降水异常等现象直接影响到畜牧业的饲草供应、畜群健康、畜产品产量和质量等方面。④ 因此，有必要将雨养农业中的畜牧部门纳入分析框架。

目前学界对于中国主要雨养农业区气候变化整体趋势的判断大致一致，即气温整体呈现上升趋势，西北雨养农业区1961~2012年平均每10年气温上升约0.3℃，2001~2010年是近半个世纪以来最暖的时段；⑤ 黄土高原地区

---

① 杨轩：《气候变化对黄土高原作物生产系统产量、水分利用及土壤养分的影响》，兰州大学2019年博士学位论文。

② 何亮：《黄土高原冬小麦物候、产量和水分利用对气候变化和波动的响应》，中国科学院大学2015年博士学位论文。

③ IPCC. IPCC Fifth Assessment Report（AR5）[M]. Cambridge：Cambridge University Press，2014.

④ Rojas-Downing, M. M., Nejadhashemi, A. P., Harrigan, T., et al. Climate change and livestock：Impacts, adaptation, and mitigation [J]. Climate Risk Management, 2017, 16：145-163.

⑤ 姚玉璧、杨金虎、肖国举等：《气候变暖对西北雨养农业及农业生态影响研究进展》，载《生态学杂志》2018年第7期。

降水减少明显，北疆以及河西走廊地区降水有所增加。① 有关模型模拟结果也与实际观测数据基本一致。然而，受限于各种气候模型本身的局限以及未来全球温室气体排放量的不确定，目前对于这些地区气候变化潜在变化幅度的判断却难以达成共识，存在较大的不确定性。

2.4　雨养农业对于气候变化的适应

对气候变化的适应指的是通过调整自然或人类系统，对实际或预期发生的气候刺激因素或他们的作用进行反应，以期减少损失或利用潜在机会。② 目前学界对在雨养农业区气候变化适应的研究多关注保护性耕作、轮作、改善肥料施用等具体的适应措施。

在许多发达国家的农业环境中，在气候变化条件下，少耕、免耕、秸秆覆盖等保护性耕作措施被认为对作物的产量和水分利用效率有显著的积极效果，保护性耕作是一种潜在的经济可行的农业气候变化适应措施。彭德尔等（Pendell et al.）提出，在美国的中部地区，即便没有任何的政策补贴，采用免耕法年均能获得每公顷 134.09～225.93 美元的收益以及每公顷 1.51～2.87tCO_2e 的固碳量。③ 格瑞斯等（Grace et al.）发现，在澳大利亚的南部地区，相较于少耕法，免耕法的利润也更多。④

---

① 刘玉洁、陈巧敏、葛全胜等：《气候变化背景下 1981～2010 中国小麦物候变化时空分异》，载《中国科学：地球科学》2018 年第 7 期。马雅丽、郭建平、赵俊芳：《晋北农牧交错带作物气候生产潜力分布特征及其对气候变化的响应》，载《生态学杂志》2019 年第 3 期。

② Stavins, R. N. The cost of carbon sequestration: a revealed-preference approach [J]. The American Economic Review, 1999, 89 (4): 994 – 1009. Hunt, C. Economy and ecology of emerging markets and credits for bio-sequestered carbon on private land in tropical Australia [J]. Ecological Economics, 2008, 66 (2): 309 – 318. IPCC. IPCC Fifth Assessment Report (AR5) [M]. Cambridge: Cambridge University Press, 2014. 侯玲玲、王金霞、黄季焜：《不同收入水平的农民对极端干旱事件的感知及其对适应措施采用的影响——基于全国 9 省农户大规模调查的实证分析》，载《农业技术经济》2016 年第 11 期。王天穷、顾海英：《我国农村能源政策以及收入水平对农户生活能源需求的影响研究》，载《自然资源学报》2017 年第 8 期。

③ Pendell, D. L., Williams, J. R., Boyles, S. B., et al. Soil carbon sequestration strategies with alternative tillage and nitrogen sources under risk [J]. Applied Economic Perspectives and Policy, 2007, 29 (2): 247 – 268.

④ Grace, P. R., Antle, J., Ogle, S., et al. Soil carbon sequestration rates and associated economic costs for farming systems of south-eastern Australia [J]. Australian Journal of Soil Research, 2010, 48: 1 – 10.

就发展中国家而言,研究发现,在印度的中央平原,稻—麦农业系统中施以免耕法并配合以 25 美元/tCO_2e 的碳减排补贴,净固碳率可达 0.3tCO_2e/公顷/年。① 在中国的黄土高原,通过在玉米—冬小麦—大豆轮作中实行长期免耕+秸秆覆盖,极大地提高了土壤持水力、作物产量和水分利用的潜力。② 相比传统耕作,实行了免耕/秸秆覆盖措施下的冬小麦产量显著提高。③ 冬小麦—春玉米的轮作序列中,实行少耕、免耕、深松的保护性耕作措施可显著提高作物产量和水分利用效率。④

还有学者发现,保护性耕作措施通过促进作物生长并提高碳输入量来提高土壤有机碳储量。对于雨养农业区而言,保护性耕作对固碳的积极效应更大,⑤ 长期应用该耕作措施,土壤固碳能力增强的同时温室气体排放通量也会减少。⑥ 耕作强度减小可对土壤剖面残留的植物体分解起到抑制作用,从而减缓土壤有机质的分解速率。⑦ 而土壤表面覆盖作物秸秆直接提高了土壤有机质来源,对浅层土壤有机碳有显著的积极作用。⑧

轮作制的经济结果在不同地区表现各异。在美国的蒙大拿州,用持续性

---

① Grace, P. R., Antle, J., Aggarwal, P. K., et al. Soil carbon sequestration and associated economic costs for farming systems of the Indo-Gangetic Plain: A meta-analysis. Agriculture, Ecosystems & Environment [J], 2012, 146 (1): 137 – 146.

② 杨轩:《气候变化对黄土高原作物生产系统产量、水分利用及土壤养分的影响》,兰州大学 2019 年博士学位论文。

③ Su, Z., Zhang, J., Wu, W., et al. Effects of conservation tillage practices on winter wheat water-use efficiency and crop yield on the Loess Plateau, China [J]. Agricultural Water Management, 2007, 87: 307 – 314.

④ Zhang, Y., Wang, R., Wang, S., et al. Effects of different sub-soiling frequencies incorporated into no-tillage systems on soil properties and crop yield in dryland wheat-maize rotation system [J]. Field Crops Research, 2017, 209: 151 – 158.

⑤ Martínez, C. C., Angás, P., Lampurlanés, J. Long-term yield and water use efficiency under various tillage systems in Mediterranean rainfed conditions [J]. Annals of Applied Biology, 2007, 150 (3): 293 – 305.

⑥ Das, T. L., Saharawat, Y. S., Bhattacharyya, R, et al. Conservation agriculture effects on crop and water productivity, profitability and soil organic carbon accumulation under a maize-wheat cropping system in the North-western Indo-Gangetic Plains [J]. Field Crops Research, 2018, 215: 222 – 231.

⑦ Franzluebbers, A. J. Achieving soil organic carbon sequestration with conservation agricultural systems in the southeastern United States [J]. Soil Science Society of America Journal, 2010, 74 (2): 347 – 357.

⑧ Liu, D. L., Chan, K. Y., Conyers, M. K. Simulation of soil organic carbon under different tillage and stubble management practices using the Rothamsted carbon model [J]. Soil & Tillage Research, 2009, 104: 65 – 73.

种植系统替代种植—休耕轮作制的成本为 4.88～56.95 美元/tCO_2e，平均固碳成本不超过 20.34 美元/tCO_2e。① 在加纳，利用玉米和花生进行轮作能够使农民的收入增长 5%～160%。② 然而在澳大利亚的西澳大利亚州，用 10 种作物进行组合的轮作制的平均成本高于 51.89 美元/tCO_2e。③ 在塞内加尔西部半干旱地区，粟—落花生轮作的固碳成本超过了 129 美元/tCO_2e。④ 在中国黄土高原雨养农业区未来气候变化下，苜蓿—小麦系统在产量、土壤养分以及经济效益方面表现出较高的适应性。在稳定浓度路径（RCP）4.5 情景⑤中，紫花苜蓿产量提高 4.0%～12.3%，苜蓿连作和苜蓿—小麦轮作的毛利润和水分效益相对历史情景提升了 0.8%～12.8%。⑥ 紫花苜蓿—玉米轮作和紫花苜蓿—高粱（*Sorghum bicolor*）轮作的水分利用效率平均增加 0.98 倍，产量平均增加 0.83 倍。⑦

学界对于通过改变施肥方式来实现雨养农业对气候变化的应对存在争论。例如，彭德尔等提出，在堪萨斯州的东北部地区，如果有足够的政策扶持，将牛粪作为肥料的种植业是经济有效的（每公顷净回报为 58.46～149.83 美元）。⑧ 阿比

---

① Antle, J. M., Capalbo, S. M., Mooney, S., et al. Economic analysis of agricultural soil carbon sequestration: an integrated assessment approach [J]. Journal of Agricultural and Resource Economics, 2001, 26 (2): 344 – 367. Antle, J. M., Capalbo, S. M., Mooney, S., et al. Sensitivity of carbon sequestration costs to soil carbon rates [J]. Environmental Pollution, 2002, 116 (3): 413 – 422.

② González-Estrada, E, Rodriguez, LC, Walen, VK, et al. Carbon sequestration and farm income in West Africa: Identifying best management practices for smallholder agricultural systems in northern Ghana [J]. Ecological Economics, 2008, 67 (3): 492 – 502.

③ Kragt, M. E., Pannell, D. J., Robertson, M. J., Thamo, T. Assessing costs of soil carbon sequestration by crop-livestock farmers in Western Australia [J]. Agricultural Systems, 2012, 112: 27 – 37.

④ Tschakert, P. The costs of soil carbon sequestration: an economic analysis for small-scale farming systems in Senegal [J]. Agricultural Systems, 2004, 81 (3): 227 – 253.

⑤ 稳定浓度路径（RCPs）是指对辐射活性气体和颗粒物排放量、浓度随时间变化的一致性预测的一个集合，其涵盖广泛的人为气候强迫。RCP4.5 的路径形状为不超过目标水平达到稳定，在 RCP4.5 情景下，辐射强迫将在 2100 年达到 4.5W/m^2，二氧化碳、甲烷和氧化亚氮的排放量将在 2040 年达到峰值，温室气体浓度将在 2070 年趋于稳定，2100 年之后稳定在 650mL·m^{-3}，到 2100 年预计升温 2.4～5.5℃（平均 3.6℃）。

⑥ 杨轩:《气候变化对黄土高原作物生产系统产量、水分利用及土壤养分的影响》，兰州大学 2019 年博士学位论文。

⑦ 毛桂莲、虎德钰、许兴等:《干旱风沙区苜蓿后茬不同轮作方式对水分利用效率和产量的影响》，载《水土保持学报》2015 年第 3 期。

⑧ Pendell, D. L., Williams, J. R., Boyles, S. B., et al. Soil carbon sequestration strategies with alternative tillage and nitrogen sources under risk [J]. Applied Economic Perspectives and Policy, 2007, 29 (2): 247 – 268.

德等（Abid et al.）指出，巴基斯坦旁遮普省北部的小农户能够通过优化施肥方式有效应对气候变化，减少潜在损失。① 然而卡克巴赞等（Khakbazan et al.）发现，在加拿大西部地区，如果借助于改变肥料施用率来实现气候变化的应对，其经济前景堪忧。②

当前学界所关注的雨养农业气候变化适应措施主要还是集中在种植部门。近年来，有科学证据显示，畜牧部门的一些措施也可以改善雨养农业对气候变化的适应。目前该部分的研究多数还是探讨相关生物物理学机理，实证研究较为缺乏。例如，有关学者指出，包括羊在内的反刍动物饲料的精粗比与其瘤胃甲烷产量之间存在紧密的联系。适量提高饲料中精料的比重利于降低动物体内乙酸的产生，从而减少甲烷的排放。同时，对于动物粪便进行表面覆盖可以减少甲烷的产生；避免粪便处于干湿交替状态还可以抑制氧化亚氮的生成与排放。③ 用苜蓿等多年生牧草代替一年生牧草可以使土壤得到有效改良，土壤中固氮量得到增加，减少农地氮肥的使用；同时，多年生牧草的水土保持能力要高于一年生作物，向多年生牧草的转化可以有效降低土壤侵蚀，减少氮的流失量。④

现有研究指出，尽管当前城市化和全球化水平不断提高，治理机能不断改善，但是广大发展中国家雨养农业地区的饥饿、贫困以及生计脆弱性问题在气候多变、气候变化风险、人口增长、生产与需求结构变化等的作用下依旧突出。⑤ 发展中国家的绝大多数贫困人口居住在雨养农业区，其生计主要

---

① Abid, M., Schilling, J., Scheffran, J., et al. Climate change vulnerability, adaptation and risk perceptions at farm level in Punjab, Pakistan [J]. Science of the Total Environment, 2016, 547: 447 – 460.

② Khakbazan, M., Mohr, R. M., Derksen, D. A., et al. Effects of alternative management practices on the economics, energy and GHG emissions of a wheat-pea cropping system in the Canadian prairies [J]. Soil and Tillage Research, 2009, 104 (1): 30 – 38.

③ 李胜利、金鑫、范学珊等：《反刍动物生产与碳减排措施》，载《动物营养学报》2010年第1期。

④ Tang, K., Hailu, A., Kragt, M. E., Ma, C. The response of broadacre mixed crop-livestock farmers to agricultural greenhouse gas abatement incentives [J]. Agricultural Systems, 2018, 160: 11 – 20.

⑤ 侯向阳、韩颖：《内蒙古典型地区牧户气候变化感知与适应的实证研究》，载《地理研究》2011年第10期。刘华民、王立新、杨劼等：《气候变化对农牧民生计影响及适应性研究——以鄂尔多斯市乌审旗为例》，载《资源科学》2012年第2期。黄泽颖、逄学思、周晓雨等：《半干旱地区马铃薯种植户适应气候变化行为研究——基于甘肃省362个微观调查数据》，载《中国农业大学学报》2019年第1期。刘玉洁、陈巧敏、葛全胜等：《气候变化背景下1981～2010中国小麦物候变化时空分异》，载《中国科学：地球科学》2018年第7期。马雅丽、郭建平、赵俊芳：《晋北农牧交错带作物气候生产潜力分布特征及其对气候变化的响应》，载《生态学杂志》2019年第3期。Cholo, T. C., Fleskens, L., Sietz, D., Peerlings, J. Land fragmentation, climate change adaptation, and food security in the Gamo Highlands of Ethiopia [J]. Agricultural Economics, 2019, 50 (1): 39 – 49.

依赖从事雨养农业生产。为了有效减少贫困,有必要从提高雨养农业生产力和气候适应性以及调整农业生产结构等方面促进发展中国家雨养农业升级转型。① 有实证研究发现,作物产量每增加1%可以使全球范围内的贫困总人数减少0.91%,而在非洲则可减少0.96%。②

已有研究认为,发展中国家雨养农业升级转型主要可以从两个方面来进行。一种是扩大从事雨养农业农户的生产规模。这在一些人口密度较低的非洲和拉美国家是可行的。然而在人口密度较高的南亚、东亚以及东非地区,由于农地利用已逼近或达到极限,进一步的扩张式粗放型雨养农业经营可能最终会加剧贫困。③ 另一种潜在途径是通过优化雨养农业生产经营结构使得其朝着高价值和商业化方向转变。即便对于许多小农户而言通过该种方式增加的收入比较有限,但该种方式对于改善粮食安全保障和提高农业的抗风险性还是具有积极意义的。④

总体而言,雨养农业对气候变化适应性措施的选择受到所考虑的具体适应策略、地理位置、农业结构、政策情景等因素的影响,存在较大差异。此外,已有研究多集中于从地区或国家等宏观层面出发,对于单一气候变化适应措施展开研究。目前基于农户微观视角,以确保农户收益优化为目的,探讨运用多种农业气候变化适应性措施及其综合影响,系统分析雨养农业生产对气候变化的适应过程和有效适应路径的相关研究较少。

---

① World Bank. Health Nutrition and Population Statistics [R]. World Bank, Washington DC, 2016. FAO, IFAD, UNICEF, WFP, WHO. The state of food security and nutrition in the world 2018: building climate resilience for food security and nutrition [R]. FAO, Rome, 2018.

② Harris, D., Orr, A. Is rainfed agriculture really a pathway from poverty? [J] Agricultural Systems, 2014, 123: 84-96.

③ Ndjeunga, J., Savadogo, K. Changes in rural household livelihood strategies and outcomes in Burkina Faso [A]. In: Freeman, H. A., Rohrbach, D. D., Ackello-Ogutu, C. (Eds.), Targeting Agricultural Research for Development in the Semi-Arid Tropics of Sub-Saharan Africa. Proceedings of a Workshop held at International Center for Research in Agroforestry [C]. Nairobi, Kenya, 2002. Banda, D. J., Hamukwala, P., Haggblade, S., Chapoto, A. Dynamic pathways into and out of poverty: A case of smallholder farmers in Zambia (Working Paper No. 56) [Z]. Zambia: Food Security Research Project, 2011. Harris, D., Orr, A. Is rainfed agriculture really a pathway from poverty? [J] Agricultural Systems, 2014, 123: 84-96.

④ Rahn, E., Liebig, T., Ghazoul, J., et al. Opportunities for sustainable intensification of coffee agro-ecosystems along an altitudinal gradient on Mt. Elgon, Uganda [J]. Agriculture, Ecosystems & Environment, 2018, 263: 31-40.

2.5 气候变化对雨养农业的影响及其适用的相关研究方法

目前学界所运用的气候变化对雨养农业的影响及其适用的相关研究方法可分为四类。

第一类是运用单纯的天气/气候模型,通过建立气候与作物产量之间的函数关系,并利用估计的生产函数模拟气候变化对农业产量的影响,可称为生产函数法。① 有关研究发现,美国东部以及伊利诺伊州自1950年以来,在气温不高于29℃的条件下,当地的玉米产量会随着气温的上升而提高;然而超过29℃的气温条件会造成玉米产量明显下降。② 在美国的艾奥瓦州、伊利诺伊州以及印第安纳州,粮食作物产量与饱和水汽压差即空气的干燥程度之间存在一定的非线性的关系。③ 在中国,尹朝静等利用粮食生产投入产出数据和气象数据实证研究发现,降水增加和气温上升均对粮食产量变动产生了正向非线性影响。降水量每增加100毫米,粮食产量增加0.67%~0.85%;温度每上升1℃,粮食产量增加0.35%~0.41%。然而,降水增加和气温上升抑制了南方地区粮食产量增长,但对北方地区却有促进作用。④ 生产函数法的一个重要缺陷在于它难以刻画农户适应气候变化的能力,在生产函数法的

---

① Kaiser, H. M., Riha, S. J., Wilks, D. S., et al. A farm-level analysis of economic and agronomic impacts of gradual climate warming [J]. American Journal of Agricultural Economics, 1993, 75 (2): 387 – 398. Lobell, D. B., Cahill, K. N., Field, C. B. Historical effects of temperature and precipitation on California crop yields [J]. Climatic Change, 2007, 81 (2): 187 – 203. 付莲莲、朱红根、周曙东:《江西省气候变化的特征及其对水稻产量的贡献——基于"气候—经济"模型》,载《长江流域资源与环境》2016年第4期。

② Schlenker, W., Roberts, M. J. Nonlinear temperature effects indicate severe damages to US crop yields under climate change [J]. Proceedings of the National Academy of sciences, 2009, 106 (37): 15594 – 15598. Roberts, M. J., Schlenker, W., Eyer, J. Agronomic weather measures in econometric models of crop yield with implications for climate change [J]. American Journal of Agricultural Economics, 2012, 95 (2): 236 – 243.

③ Lobell, D. B., Hammer, G. L., McLean, G., et al. The critical role of extreme heat for maize production in the United States [J]. Nature Climate Change, 2013, 3 (5): 497 – 501. Lobell, D. B., Roberts, M. J., Schlenker, W., et al. Greater sensitivity to drought accompanies maize yield increase in the US Midwest [J]. Science, 2014, 344 (6183): 516 – 519.

④ 尹朝静、李谷成、高雪:《气候变化对中国粮食产量的影响—基于省级面板数据的实证》,载《干旱区资源与环境》2016年第6期。

基本设定下，农户既不能引入新的作物品种，也无法改变农地用途。

第二类是运用特征（Hedonic）农地价格模型，并将天气因素纳入模型中作为影响农地价格的关键因素之一。该方法以农地价格作为被解释变量，解释变量包含气候因素和其他一些反映农地价格的特征。① 在一些情况下，该方法估计结果得出了与生产函数法截然不同甚至相反的结论，即气候变化对农业的负面影响非常小，甚至还可能存在正面影响。该类模型的前提是农地交易市场处于均衡当中。考虑到地理因素异质性、农地交易频率偏低以及农户对土地价格的估计易受外部因素影响等原因，利用该类方法的有关研究所得出的结论往往存在较大差异。② 此外，农地价格还受到一些无法衡量的农地质量因素的影响，而这些因素与天气因素之间有时存在关联性。在此种情况下，有关气候变化对于农地价格影响作用的解释容易产生混淆。此外，该方法也可能忽视减缓措施的关键作用。值得注意的是，一些运用该类方法的研究只涵盖了横截面数据分析，因而其所得到的有关结论的有效性值得商榷。

第三类是将农户生产的利润与天气因素进行关联。例如，有学者利用县一级的横截面数据分析发现，天气的波动会反映到农场年均单位面积利润上，并最终会反映到农地价值上。德谢勒斯和格林斯通（Deschenes and Greenstone）利用1987年、1992年、1997年和2002年的全美农业调查所提供的县级横截面数据分析发现，天气变化对美国农地价值的初步影响范围为-2020亿~3210亿美元，约占全美农地总价值的-18%~29%；进一步改进分析方法的结果显示，天气的变化会使全美农地价值增加952亿~1108亿美元。③ 费雪等（Fisher et al.）利用德谢勒斯和格林斯通的数据，并对相关变量数值进行了进一步完善，在此基础上进行了类似的分析。他们发现，德谢勒斯和

---

① Mendelsohn, R., Nordhaus, W. D., Shaw, D. The impact of global warming on agriculture: a Ricardian analysis [J]. The American economic review, 1994: 753 – 771. Schlenker, W., Hanemann, W. M., Fisher, A. C. Will US agriculture really benefit from global warming? Accounting for irrigation in the hedonic approach [J]. The American Economic Review, 2005, 95 (1): 395 – 406.

② Amundson, R., Guo, Y., Gong, P. Soil diversity and land use in the United States [J]. Ecosystems, 2003, 6 (5): 470 – 482. Huffman, W. E., Jin, Y., Xu, Z. The economic impacts of technology and climate change: new evidence from US corn yields [J]. Agricultural Economics, 2018, 49 (4): 463 – 479.

③ Deschenes, O., Greenstone, M. The economic impacts of climate change: evidence from agricultural output and random fluctuations in weather [J]. American Economic Review, 2007, 97 (1): 354 – 385.

格林斯通低估了天气变化对美国农地价值的影响，证实了气候变化对农业的负向影响依然成立。① 冯晓龙等利用陕西八个县的实地调查数据分析了农户气候变化适应性决策对农业产出与产出风险的影响效应。他们发现，采取适应措施的农户若未采取相应的适应措施，其亩均产出将下降 3.7%，产出风险将增加 107%；未采取适应措施的农户若采取相应的适应措施，其亩均产出将增加 5.5%，产出风险将下降 19.6%。② 除了气温和降雨之外，其他天气因素，如相对湿度、风速、日照时间和蒸发率等也可能会影响农业产出。还有学者的研究强调了模型中控制这些天气变量的重要性，利用 1980~2010 年中国县级层面农业产出和站点气象统计数据，发现忽视相对湿度这一因素的积极贡献将导致气候变化（增温）的负效应被严重高估，其中，对小麦产量的影响高估了 29.6%。③ 然而，这一类的分析没有将农机和劳动力使用成本纳入生产成本的核算当中，也忽略了技术进步对于农户利润的影响。

第四类是将实际长期观测或实验数据与作物模型相结合，分析气候变化情景下作物产量、蒸散分配、水分利用、土壤养分和经济效益的变化，明确作物生产受不同气候变化影响的机制，评估作物生产的适应性。何亮分析了黄土高原气象数据、农业气象试验站观测数据、长尺度产量、水分管理等数据，发现黄土高原区域的冬小麦播种、出苗、越冬开始分别平均推迟了 1.2、1.3、1.2 天/10 年；返青、开花、成熟期平均分别提前了 2.0、3.7、3.1 天/10 年；生育期、越冬期（越冬开始到返青开始）和出苗到开花（营养生长）平均分别缩短了 4.3、3.1 和 5.0 天/10 年，而开花到成熟（生殖生长）延长了 0.7 天/10 年。④ 杨轩结合大田试验数据和作物模型 APSIM，依据不同气候变化情景，对黄土高原雨养农业长期尺度下传统作物系统和改进的作物系统的生产进行了模拟。结果显示，在气温升高、降水量减少的情景下，冬小麦、玉米和紫花苜蓿产量降低，最大幅度分别可达 38.7%、40.3% 和 41.8%。冬小麦和紫花苜蓿在气温降低、降水量增加时增产，最大增幅分别为 29.8% 和

---

① Fisher, A. C., Hanemann, W. M., Roberts, M. J., et al. The economic impacts of climate change: evidence from agricultural output and random fluctuations in weather: comment [J]. American Economic Review, 2012, 102 (7): 3749-3760.
② 冯晓龙、刘明月、霍学喜等：《农户气候变化适应性决策对农业产出的影响效应——以陕西苹果种植户为例》，载《中国农村经济》2017 年第 3 期。
③ 李承政、顾海英：《气候冲击对中国县级经济的影响研究》，格致出版社 2019 年版。
④ 何亮：《黄土高原冬小麦物候、产量和水分利用对气候变化和波动的响应》，中国科学院大学 2015 年博士学位论文。

51.7%。玉米在降水量增加、温度不变的情景下增产幅度最大，为22.0%。在设定范围内，紫花苜蓿的产量变异范围受气候变化的影响最小，说明其适应能力较强。① 扎莫（Thamo）将长期观测数据与APSIM、GrassGro与MIDAS模型模拟相结合，认为在澳大利亚中部谷物带（Central Wheatbelt），在适当适应行为的干预下，作物产量受气候变化的影响可以控制在一定的范围之内。②

2.6 国内外相关文献的总结性评述

气候变化下的雨养农业生产是一个涉及多学科的研究领域，引起了不同学科背景学者的关注，也是近年来开始兴起的一个研究热点。已有研究对于气候变化条件下的雨养农业生产问题进行了多角度的机理剖析和实证检验，丰富了人们对于雨养农业生产与气候变化之间关系的认识，为应对气候变化农业政策的制定与实践提供了有价值的参考。然而，该领域在以下方面亟待进一步拓展和完善。

第一，学科隔阂较为明显，评估存在偏差。气候变化对雨养农业生产影响机理以及雨养农业对气候变化适应的研究还存在较为明显的学科隔阂，未能将自然科学领域的气候变化模型、作物生长模型、畜牧生产模型等与微观层面的农户应对气候变化的行为、宏观层面的农业政策等经济社会因素结合起来，导致当前气候变化对雨养农业影响以及雨养农业对气候变化适应的评估存在偏差。

第二，现有研究多关注发达国家种植业，而关注气候变化对于雨养农业区畜牧业影响的研究较少，对发展中国家雨养农业在气候变化所引起的自然环境与制度环境方面影响的综合作用下的种—畜结构、粮食作物与经济作物比例、农地分配、农户收益、农业温室气体排放等问题鲜有涉及。目前尚缺乏对不同类型雨养农业生产系统进行相关问题分析的国际比较研究。

第三，有关雨养农业对气候变化的适应方面的研究多集中于从地区或国

---

① 杨轩：《气候变化对黄土高原作物生产系统产量、水分利用及土壤养分的影响》，兰州大学2019年博士学位论文。

② Thamo, T. Climate change in Western Australia agriculture: A bioeconomic and policy analysis [D]. Ph. D. thesis, School of Agricultural and Resource Economics, University of Western Australia, 2017.

家等宏观层面出发对于单一气候变化适应措施进行探讨。目前基于农户微观视角，以确保农户收益优化为目的，探讨运用多种农业气候变化适应性措施及其综合影响，系统分析雨养农业生产对气候变化的适应过程和有效适应路径的相关研究较少。

第四，现有相关研究所利用的研究方法多建立在雨养农业的某一单一模块基础上，对于各模块之间的相互关系以及共同作用涉及较少。目前尚缺乏基于雨养农业生产系统整体，综合考量生物、环境、管理、技术以及财务等各方面条件的综合分析工具。

针对当前研究的不足，本书尝试对其做进一步充实和完善。

第3章 理论基础

3.1 农业系统理论[*]

农业是指通过对于植物、动物、菌类以及其他类型生命体进行培养，从而为人类的生存和发展提供食品、纤维、生物燃料、药物以及其他产品的产业。农业系统是一个由农业生态环境系统和农业经济社会系统耦合而成的多层次的复杂巨系统，包含人口、环境、资源、经济和社会五类要素（见图3-1）。[①] 农业生态系统是农业的本体系统，提供了农业赖以发展的生物物理平台。而农业经济社会系统是农业的能动系统，提供了农业存在与发展所需的动力。农业生态系统与农业经济社会系统之间依据自身的运行规律，在复合系统内进行不同物质、能量和信息的交替、转化，从而形成新的功能体。这一新的功能体所产生的物质、能量以及信息具有新的特征和独立的结构，有别于单独的农业生态系统与农业经济社会系统。农业系统是一个复杂的灰色系统，存在众多要素的作用关系和过程。[②]

整体性是农业系统的重要特征。农业系统内的各要素相互关联、相互制约，在系统内形成了错综复杂的关系网络。物质、能量和信息在农业生态环境系统内部、农业经济社会系统内部以及以上两系统之间不断合成、分解和

---

[*] 该节部分内容参考唐凯：《基于生物经济学的澳大利亚农业温室气体减排潜能分析》，人民出版社2018年版。

[①] 王芳：《西部循环型农业发展的理论分析与实证研究》，华中农业大学2006年博士学位论文。

[②] 王继军、姜志德、连坡等：《70年来陕西省纸坊沟流域农业生态经济系统耦合态势》，载《生态学报》2009年第9期。李玥：《黄土丘陵区退耕与农业生态经济社会系统协同发展研究——以安塞县为例》，西北农林科技大学2019年博士学位论文。

流动，是形成农业系统结构与功能的物质内容。农业系统内的关系网络与物质内容一起形成了农业系统的功能与性质。农业系统内部是一种开放、非平衡、非线性的相互作用关系。

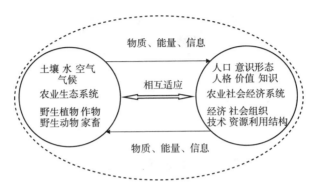

图 3-1　农业系统示意图

基于一般系统理论与自组织理论的分析，从微观层面进一步对系统内部演化的分析得到了复杂适应系统理论。这一理论揭示了系统内部各个单元和要素具有相适宜的特性，系统内部的物质传递和信息交换正是由于这一特性不断驱动系统的演化。同时，由于该理论的出发角度是极其微观的，使其可以明确各个单元和要素的演替过程，剖析系统演变的本质问题，而不仅受限于对系统的简单阐述，因此，该理论的产生可以为研究系统演化过程中出现的多种问题给予一定程度的解释与帮助。雨养农业系统是由多个子系统及其内部多种单元和要素共同作用所形成的一个复杂适应系统，随着系统外界的变化，对这一新的功能体内的各个子系统及内部单元产生了作用，内部的各个子系统及其各类要素之间开始频繁地交换信息与能量，逐步突破原有的时间与空间限制的状态，促使复合系统的更新。在外界作用下，雨养农业系统内部自我革新的能力不断调整、演化、优化。

系统在稳定状态与不稳定状态之间的交替是随着外界的影响和干扰而改变的，这种交替的变化有时候就发生在一瞬间，以此理论为依据，可以用来认识和预测复杂的系统行为。气候变化与雨养农业系统的突变是各个子系统构成要素共同突变的结果。生态系统的构成要素主要是环境和资源，经济社会系统的要素则是产业和效益以及人的认知及感受等。在气候变化发生前，雨养农业系统的基本构成要素所构成的生态、经济、社会环境在一定时期内处于相对稳定的状态；随着气候变化的发生以及雨养农业区域外部环境发生的重大变化，雨养农业系统内的各类要素突变引起了气候变化与雨养农业系统的整体突变，原有稳定链网结构的破坏，形成了新的功能结构体，在新的结构下气候变化与雨养农业系统之间重新相互作用，改变了原有的作用路径、方向及强度。①

从系统的角度来看，将农业系统分解为部分而进行单独研究，会切断该部分与其他部分之间的关联关系，并不能回答农业系统整体性的问题。具体到雨养农业，系统理论就要求从雨养农业系统整体的视角出发，对采取了气候变化适应措施的雨养农业系统所包含的经济规律进行探讨，着眼研究气候因素的变化以及随之产生的雨养农业政策体系对于雨养农业系统整体运行的影响，共同探索雨养农业在气候变化条件下的发展规律，促进雨养农业生态环境系统和雨养农业社会经济系统的协调演进，以期实现雨养农业的持续健康发展。

本书将此理论作为运用生产者微观视角，系统地分析气候变化对雨养农业生产的影响，评价雨养农业对气候变化的环境效应和适应成本，探索在优化农户收益的前提下雨养农业对气候变化的有效适应路径的理论依据。

3.2 外部性理论

讨论雨养农业对气候变化的适应以及环境规制的经济学本质，有必要回顾环境经济学最核心的概念——外部性（externality）。庇古（Pigou）率先提出了运用征税来治理污染的方法，将外部性理论系统化，使之成为新古典经济学的核心内容之一。② 科斯（Coase）进一步提出了利用市场解决外部性内部化的方法，③ 自此，外部性理论成为新制度经济学的重要组成部分。

依据外部性理论，市场机制在环境资源配置方面由于外部性的存在而产生失灵。所谓外部性，可以理解为经济个体的行为对其他经济个体的福利造成了影响，却不会为之付出代价或得到补偿。外部性的实质在于，这种外部影响只能由非价格机制传递而不能通过市场价格进行买卖。依据对其他个体

---

① 李玥：《黄土丘陵区退耕与农业生态经济社会系统协同发展研究——以安塞县为例》，西北农林科技大学2019年博士学位论文。
② Pigou, A. The Economics of Welfare [M]. London: The Macmillan Company, 1920.
③ Coase, R. H. The problem of social cost [J]. The Journal of Law and Economics, 1960, 3: 1-44.

的福利所造成影响的效用来看,外部性分为正外部性(外部经济)和负外部性(外部不经济)。正外部性概念源于马歇尔提出的"外部经济",而负外部性概念源于庇古的"外部不经济"概念。当外部性为其他个体带来有益影响时,此时的外部性称为正外部性;反之,当影响为有害时,称为负外部性。

环境问题的产生与外部性紧密相关:经济主体从事正外部性的经济活动而不能获得相应的利益补偿,积极性受挫;从事负外部性的经济活动不必为此付出相应的代价,社会承担损失导致外部负效应迅速扩展,环境污染不断加剧,生态环境遭到破坏。环境管制政策工具的演变实际上是随着人们对外部性的认识深化而不断演进的,具体表现为:对外部性的认识经由私人、社会成本的背离—产权界定不清晰—交易费用过高等不断深化的认识过程。①

外部性问题是农业温室气体排放的经济学本质。作为经济系统中的个体,农业部门中产生温室气体排放的活动属于私人行为。然而,由温室气体排放所带来的负面影响却会波及所有的个体。私人决策没有考虑到由此产生的外部成本,此时的农业温室气体排放所带来的边际社会成本(如图3-2中的 MSC)高于边际私人成本(如图3-2中的 MPC),产生负外部性。

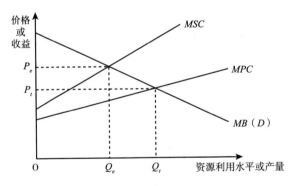

图3-2 农业温室气体排放的负外部性

外部性问题同样是农业温室气体减排的经济学本质。农业部门减排温室气体所带来的收益能够由所有人共享,由此带来了外部收益。当私人收益不包括该部分外部收益时,边际社会收益高于边际私人收益,产生正外部性。

环境资源的公共产品属性、环境污染的负外部性和微观经济主体的有限理性,决定了市场机制自身在解决环境问题时会出现市场失灵的状况,这需

---

① 薛豫南:《基于循环经济的畜禽污染治理动力机制》,大连海事大学2020年博士学位论文。

要政府的有效干预。环境规制是社会性规制的重要内容之一，它的特殊性在于政府制定一系列的经济、行政以及其他政策或措施，例如，通过征收排污费或界定环境资源产权并对环境资源合理定价等来调节企业的生产活动，使得企业在进行生产决策时把外部成本考虑进来，实现边际私人成本和社会成本相一致，进而解决环境污染负外部性，提高资源配置效率，实现环境质量改善与经济发展协调目标。[①]

依据外部性理论，实行碳汇农业在经济学上是从农业温室气体排放的负外部性转向农业温室气体减排[②]的正外部性。此过程中的关键是实现外部性的内部化。此内部化过程可以通过命令—控制型规制（command-and-control regulations）、市场型规制（market-based instruments）以及自愿型规制（volunteer regulations）来实现。[③]

命令—控制型规制可以通过立法程序对农业温室气体的排放等行为进行强制管控，也可以通过行政权力对农业温室气体的排放、监测等实施管理。通常可以采用环保立法、温室气体排放标准、温室气体排放许可等手段，对农业温室气体排放实施强制管控，体现出很强的制度刚性。

市场型规制主要通过环境税（费）、环境补贴（返还）、环境罚款、合同—契约机制、补偿规则机制、排污许可交易、抵押—返还制度、环保信贷优惠、环保差别税率等措施，对生产经营主体的成本收益结构进行调整，进而影响其决策和行为，达到减少温室气体排放、减轻气候变化后果等政策目标。

自愿型规制是行业、企业或其他经济主体提出的不具备强制约束性的约定，包括各类协定、规划和承诺，受规制对象是否参与不做强制规定，其目标是降低环境污染、保护环境质量。[④]

命令—控制型规制与市场型规制相比，本质区别在于，前者的作用机制并不直接通过市场机制传导，而更多地体现为法律法规和行政力量对污染行为的直接性强制措施；后者虽然可能也是通过政府出台相应的管制政策，但

---

① 傅京燕：《环境规制与产业国际竞争力》，经济科学出版社2006年版。
② 此处的农业温室气体减排具体描述的是一个农业系统在采取减排措施后，相较于采取措施前，温室气体的排放量减少，或\且温室气体的吸收量增加。此时对社会的正外部性来自减少的排放量与增加的吸收量之和。
③ Aldy, J. E., Stavins, R. N. The promise and problems of pricing carbon: theory and experience [J]. Journal of Environment & Development, 2012, 21 (2): 152 - 180.
④ 彭佳颖：《市场激励型环境规制对企业竞争力的影响研究》，湖南大学2019年博士学位论文。

主要是通过市场价格机制对被规制者的决策行为进行干预。

自愿型规制是规制农业温室气体排放既有政策工具体系的重要补充,但前提是有一个高度成熟的社会和一个现代治理的政府。公众参与型环境规制政策,在信息不对称、逆向选择、道德风险的情况下,可以有效解决农业温室气体排放和污染行为难以监测的问题,具有制度比较优势。①

通过推动雨养农业对气候变化的适应来实现正外部性的目的,一般会为农民带来额外的成本。基于理性经济人的假设,可以推断,如果不能因其采取农业减排措施而得到激励或补偿,农民往往会缺乏持续主动进行适应的动力和积极性。为了推出相应的激励或补偿措施,对于相关成本的准确估计就显得非常重要。

政府为了内化农业温室气体排放所关联的外部性往往会提出不同的减排激励计划。农民在思考是否参与这些减排激励计划时,会权衡参与计划后的成本与收益。只有当成本小于收益时,减排激励计划才会对农民有吸引力;反之,减排计划将难以在农民中推广。因此,评估不同激励计划所引起农业系统成本与收益的影响是十分必要的。

综上所述,本书将探讨对实现向正外部性转化以及内化外部性有着重要影响的农业温室气体减排成本,以及不同农业温室气体减排政策方法的成本与收益。

3.3 公共物品与产权理论

公共物品理论是公共经济学的主要理论之一。公共物品供给是当代政府的基本职能,政府经济行为应通过合理配置公共物品资源来展开。现代公共物品理论的诞生,是以萨缪尔森(Samuelson)在1954年发表的《公共支出的纯理论》为标志,首次将公共物品与帕累托效率联系起来,并给出了公共物品有效提供的边际条件。② 随着公共物品理论的发展,不同的理论学派对公共物品理论均提出了相应的观点,并从不同视角就公共物品所引发的经济学问题展开讨论。

---

① 周志波:《环境税规制农业面源污染研究》,西南大学2019年博士学位论文。
② Samuelson, P. A. The pure theory of public expenditure [J]. The Review of Economics and Statistics, 1954, 36 (4): 387-389.

当受益者人数众多且排除任何一个在技术上不可行的受益者时，该物品就可视为公共物品。公共物品的存在意味着个人能够持免费搭车的态度任意使用。环境就其本身固有的属性来说，具有公共性和非排他性，任何人都不能把环境资源视为自己的私有财产，而任何人又不能排除在对环境资源的享用之外。市场主体通常只从自身的角度考虑所面临的各种选择的成本和收益，将环境作为没有价值的产品对待，而将经济活动过程中的环境成本转嫁给社会。由于环境资源和环境保护的公共物品性质，非竞争性的存在导致消费者只顾及当下利益，在以各自利益为核心的行为模式下，对环境资源产生不合理的过度消费，生态环境资源遭到破坏，导致生态环境资源稀缺，最终造成社会上全体成员的利益都遭到损失。非排他性造成保护环境的行为不能同时得到生态效益与经济效益，因此，在缺少激励措施的情况下，全体社会成员都不愿意对保护环境的行为进行支付，公共产品不能足额供给。①

从经济学的角度看，环境已成为稀缺资源，稀缺意味着对一定的物品存在竞争使用，其需求不一定得到满足。环境作为公共物品，对环境的使用提出了一个配置的问题。公共物品理论作为当代环境经济的理论支柱之一，其发展脉络折射出了学界对"公共物品"供给的认知变迁历程：由新古典范式强调的公共物品消费环节上的相互依赖（非竞争性和非排他性），逐渐发展到交易范式强调的公共物品决策过程中的相互依赖（集体决策中的相互依赖才是公共物品的本质特征）。② 农业应对气候变化不具备良好的排他性，排他成本问题成为农业面源污染"非排他性"的延续。排他成本过高也是引起地方政府环境规制竞争的一个重要因素。

产权是指物品所有权，是经济所有制在法律中的体现。产权的明确界定是市场交易顺利进行的基础与保障，而市场交易的本质也就是对产权的交易。产权理论起源于对公共物品外部性的讨论，1960 年，科斯在《社会成本问题》中提出了产权理论。③ 产权理论认为，产权不明晰是外部性产生的一个典型来源。与庇古为代表的福利经济学家们所持观点不同，科斯等学者认为，外部性是相互的而非单向的，当企业产生的环境污染给居民带来损害，企业给予居民赔偿时，可能会导致居民过度的"受害者行为"，进而导致经济无效率。企业与居民责任确定、外部性损害与收益分配以及政府管制成本均是

---

① 薛豫南：《基于循环经济的畜禽污染治理动力机制》，大连海事大学 2020 年博士学位论文。
② 秦天：《环境分权、环境规制与农业面源污染》，西南大学 2020 年博士学位论文。
③ Coase, R. H. The problem of social cost [J]. The Journal of Law and Economics, 1960, 3: 1–44.

庇古的外部性理论无法解决的。外部性问题的根源不是在于市场机制本身的缺陷，而是在于对公共物品缺乏清晰的产权定界，导致市场交易机制在资源配置上丧失了有效性。在产权明晰的前提下市场机制才能够发挥正常作用，且产权必须具备排他性和可转让性。但由于环境本身具有不可分割的特性，无法界定其产权或界定成本很高，导致无法明晰产权，结果人们为了追求个人利益的最大化而无节制地争夺有限的环境资源，从而导致环境质量日益恶化。

为此，科斯提出，在交易费用的约束下，可通过界定产权的方式处理外部性。在产权明晰的前提下，若交易成本很低或为零，不管初始产权如何分配，理性的经济主体在追求利益最大化的同时会自发将外部性成本和收益考虑在内，通过双方磋商便可以保证私人成本和社会成本的一致性，从而实现资源的有效配置。若交易成本大于零，可以通过企业之间的兼并或者政府干预，通过政策制定的成本—收益分析，实现外部性问题内部化。根据科斯产权理论，环境污染等负外部性问题的产生是由于环境资源产权不明晰，因此，可以通过界定产权或市场交易来解决外部性问题，而政府需要确定初始环境资源产权的分配方式并为市场交易创造条件。现实中各个国家推行的排污许可证和排污权交易制度均是根据科斯产权理论而制定的环境政策。

由科斯提出的产权理论不仅是制度经济学的基础，更是运用除了碳税减排以外的另一种经济手段解决碳减排问题的有效方案，促使碳排放权交易与科斯产权理论之间的协调统一。碳排放权又称碳权，是指企业向大气中排放以二氧化碳为主的温室气体的权利，也是由政府人为制定排放主体在一定时期内可排放一定总量温室气体的许可。这是由于碳排放权不仅具有常规商品属性，而且其产生于人类对大气环境保护意识的崛起，人们将大气环境这种公共物品通过人为手段进行私有化，使其作为一种特殊商品，具有稀缺性、排放性、强制性、波动性、政策性等多重特征，政府允许排污市场主体对自身获得的碳排放权指标进行包括占有、买卖交易、转让使用等一系列的自由支配活动，通过上述自发的碳排放权交易以实现碳排放权的最优配置，资源优化配置目标的实现过程便构成了碳交易的基本雏形。[①] 按照科斯产权理论，碳排放权基于排污权交易理论已成为一种特殊商品，尽管对其产权的界定十分不易，但多年来碳排放权的产权界定也在发展中前进，各国减排的行动不

---

① 孙悦：《欧盟碳排放权交易体系及其价格机制研究》，吉林大学2018年博士学位论文。

断深化。对于市场主体而言,碳排放权也是一种有价值的资产。

政府作为这种商品的所有者,第一步便是在既定减排目标的约束下,向减排企业分配碳排放配额,明确其各自拥有的碳排放权产权,政府运行法律、监管等综合手段促进公共物品商品化过程,确保产权清晰,减少资产被无偿占用的可能是碳排放交易实现帕累托最优的基础。而在政府充分发挥其职能后,依靠完善的碳排放交易市场体系,交易主体之间可根据生产边际成本与减排边际成本之间的差值对比自主决策,针对配额开展购买、出售等一系列的交易活动,使碳排放问题的外部性以内部化的市场方式解决,为平衡企业边际收益与社会边际收益的共同最大化提供一条新的解决路径。[①]

环境产权在对资源的合理配置中主要解决两个问题,一是如何消除利用环境资源过程中的外部性;二是如何对环境资源进行合理定价。对于外部性问题,人们一直强调国家干预的作用,直到科斯理论产生后,人们才意识到,在应对因外部性而造成的市场失灵问题上,市场机制并非无能为力,只要产权界定清晰,交易费用控制在一定的范围内,依靠当事人之间的自由交易,就可以实现有效率的结果。

---

① 孙悦:《欧盟碳排放权交易体系及其价格机制研究》,吉林大学 2018 年博士学位论文。

第4章 澳大利亚雨养农业温室气体的减排潜力和成本

4.1 引　言

对全球而言，如何有效地减少人类温室气体排放是应对气候变化的关键。① 农业排放了约占全球12%的温室气体（年排放量超过61亿tCO_2e），其中包括全球约60%的氧化亚氮和约50%的甲烷。② 雨养农业作为全球最大的农业生产系统之一，是全球超过1/3人口的主要收入来源，产出了全球大部分的粮食。③ 现有研究显示，全球雨养农业年均温室气体排放量超过30亿tCO_2e，约占全球农业温室气体排放的1/2。因此，在包括澳大利亚在内的许多国家，雨养农业所产生的温室气体排放亟待得到有效控制。④ 考虑到农业部门在温室气体减排进程中的重要角色，雨养农业理应为全球减排努力作出

---

① Gong, C., Tang, K., Zhu, K., Hailu, A. An optimal time-of-use pricing for urban gas: A study with a multi-agent evolutionary game-theoretic perspective [J]. Applied Energy, 2016, 163: 283 – 294. Yang, L., Tang, K., Wang, Z., An, H., Fang, W. Regional eco-efficiency and pollutants' marginal abatement costs in China: A parametric approach [J]. Journal of Cleaner Production, 2017, 167: 619 – 629. Tang, K., Hailu, A. Smallholder farms' adaptation to the impacts of climate change: Evidence from China's Loess Plateau [J]. Land Use Policy, 2020, 91, 104353.

② Smith, P. Agricultural greenhouse gas mitigation potential globally, in Europe and in the UK: What have we learnt in the last 20 years? [J]. Global Change Biology, 2012, 18 (1): 35 – 43.

③ Maestre, F. T., Quero, J. L. Gotelli, N. J. Plant species richness and ecosystem multifunctionality in global drylands [J]. Science, 2012, 335 (6065): 214 – 218.

④ Kragt, M. E., Pannell, D. J., Robertson, M. J., Thamo, T. Assessing costs of soil carbon sequestration by crop-livestock farmers in Western Australia [J]. Agricultural Systems, 2012, 112: 27 – 37. Tang, K., Hailu, A., Kragt, M. E. Ma, C. The response of broadacre mixed crop-livestock farmers to agricultural greenhouse gas abatement incentives [J]. AgriculturalSystems, 2018, 160: 11 – 20.

应有贡献。

近年来,一些国家开始出台旨在减少农业温室气体的政策。总体而言,这些政策使用市场型环境规制来鼓励从事雨养农业生产的农民积极采取温室气体减排行为。① 然而,这些政策往往缺乏清晰有针对性的操作指导。对政策制定者而言,设计和实施细致有效的雨养农业温室气体减排项目存在诸多困难。其中,一项基础而必不可少的工作就是评估雨养农业温室气体减排的成本有效性。现有文献对这一重要问题却鲜有涉及。本章将对这一问题进行研究。

为了充分理解雨养农业温室气体减排的成本有效性,评估其边际减排成本和减排潜力是十分基础且重要的步骤。污染减排的最小成本理论认为,为了实现全社会的有效减排,每一个生产部门都应当取得相同的边际减排成本。② 考虑到不同生产部门有着不同的成本特征,那些边际减排成本较低而减排潜力较大的生产部门的减排机会成本更低,因此,在这些生产部门进行减排的成本有效性更强。③ 在实践中,社会将从要求减排成本低、潜力大的生产部门减少更多的温室气体排放中受益。④ 因此,政策制定者需要了解雨养农业温室气体减排的边际减排成本以及减排潜力,以找到雨养农业生产部门低成本且有效的减排方法。此外,评估边际减排成本和减排潜力还是政策制定者在设计排放税以及跨生产部门排放权交易市场时的重要参考。⑤ 例如,

---

① European Commission. 20 20 by 2020: Europe's Climate Change Opportunity COM (2008) 30 final Commission of the European Communities [R]. Brussels, Belgium, 2008. Australian Government. Emissions Reduction Fund Green Paper [R]. Commonwealth of Australia, Canberra, 2014.

② Färe, R., Grosskopf, S., Weber, W. L. Shadow prices and pollution costs in US agriculture [J]. Ecological Economics, 2006, 56 (1): 89 – 103.

③ Färe, R., Grosskopf, S., Weber, W. L. Shadow prices and pollution costs in US agriculture [J]. Ecological Economics, 2006, 56 (1): 89 – 103. Tang, K., Hailu, A., Yang, Y. Agricultural chemical oxygen demand mitigation under various policies in China: A scenario analysis [J]. Journal of Cleaner Production, 2020, 250, 119513. Adenuga, A. H., Davis, J., Hutchinson, G., Patton, M., Donnellan, T. Modelling environmental technical efficiency and phosphorus pollution abatement cost in dairy farms [J]. Science of the Total Environment, 2020, 714, 136690.

④ Tang, K., Hailu, A., Kragt, M. E., Ma, C. Marginal abatement costs of greenhouse gas emissions: broadacre farming in the Great Southern Region of Western Australia [J]. Australian Journal of Agricultural and Resource Economics, 2016, 60 (3): 459 – 475.

⑤ Färe, R., Grosskopf, S., Weber, W. L. Shadow prices and pollution costs in US agriculture [J]. Ecological Economics, 2006, 56 (1): 89 – 103. Tang, K., He, C., Ma, C., Wang, D. Does carbon farming provide a cost-effective option to mitigate GHG emissions? Evidence from China [J]. Australian Journal of Agricultural and Resource Economics, 2019, 63 (3): 575 – 592.

参照澳大利亚减少排放量基金的实际操作过程，边际减排成本可以作为政府设计碳排放权拍卖交易计划时的参考价格。只有减排成本低于该边际减排成本的投标才会被纳入碳排放权拍卖中。也就是说，边际减排成本可以作为支付给每吨减少的温室气体补贴的上限。

通过使用距离函数分析法，研究者能够在缺乏市场信息的情况下测算出生产中所产生的厌恶产出（undesirable output）[1]（例如污染物）的边际减排成本。[2] 利用该方法，需要在环境敏感性生产中识别出污染物。污染物的边际减排成本（影子价格）可以从满足对偶关系的一个距离函数中推导出。[3] 此外，距离函数分析对数据没有过于严格的要求，在应用时较为灵活，生产函数的参数型过程有多种，且有多种求解方法（例如，常用的有非参数型DEA法和参数型法）。因此，近年来距离函数分析法被全球越来越多的研究者所采用，主要用于分析与环境相关的生产技术问题。

在农业领域，距离函数分析法已经被运用于分析农业投资、生产力、生产技术发展、农地使用绿色效率、生态效率以及作物多样化等问题。[4] 一些学者也开始尝试利用距离函数分析法来评价雨养农业的生产效率。维拉诺等（Villano et al.）发现，新南威尔士州的雨养种—畜复合农业生产存在着无效

---

[1] 也称作非期望产出、坏产出。

[2] Färe, R., Grosskopf, S., Weber, W. L. Shadow prices and pollution costs in US agriculture [J]. Ecological Economics, 2006, 56 (1): 89 – 103. Tang, K., Hailu, A., Kragt, M. E., Ma, C. Marginal abatement costs of greenhouse gas emissions: broadacre farming in the Great Southern Region of Western Australia [J]. Australian Journal of Agricultural and Resource Economics, 2016, 60 (3): 459 – 475. Tang, K., Gong, C., Wang, D. Reduction potential, shadow prices, and pollution costs of agricultural pollutants in China [J]. Science of the Total Environment, 2016, 541: 42 – 50. Wei, X., Zhang, N. The shadow prices of CO_2 and SO_2 for Chinese Coal-fired Power Plants: A partial frontier approach [J]. Energy Economics, 2020, 85, 104576.

[3] Liu, H., Owens, K. A., Yang, K., Zhang, C. Pollution abatement costs and technical changes under different environmental regulations [J]. China Economic Review, 2020, 62, 101497.

[4] Blancard, S., Boussemart, J. P., Briec, W., Kerstens, K. Short-and long-run credit constraints in French agriculture: A directional distance function framework using expenditure-constrained profit functions [J]. American Journal of Agricultural Economics, 2006, 88 (2): 351 – 364. Brümmer, B., Glauben, T., Lu, W. Policy reform and productivity change in Chinese agriculture: A distance function approach [J]. Journal of Development Economics, 2006, 81 (1): 61 – 79. Picazo-Tadeo, A. J., Beltrán-Esteve, M., Gómez-Limón, J. A. Assessing eco-efficiency with directional distance functions [J]. European Journal of Operational Research, 2012, 220 (3): 798 – 809. Nguyen, H. Q. Analyzing the economies of crop diversification in rural Vietnam using an input distance function [J]. Agricultural Systems, 2017, 153: 148 – 156. Xie, H., Chen, Q., Wang, W., He, Y. Analyzing the green efficiency of arable land use in China [J]. Technological Forecasting and Social Change, 2018, 133: 15 – 28.

率，这些农民能够在不增加投入的情况下增加农产品产出。① 然而，目前利用距离函数分析法来评价雨养农业生产效率问题的实证研究还较少。

许多学者对包括杀虫剂、除草剂和悬浮沉积物在内的澳大利亚农业污染物的减排问题进行了研究。② 然而，对于澳大利亚雨养农业系统中温室气体的减排潜力和影子价格的分析还较为少见。一些研究利用田野实验或生物物理学模拟，对不同雨养农业区域的土壤碳固定潜力进行了测算。研究结果显示，对于不同的雨养农业区域，其土壤碳固定潜力具有一定的差异。③ 目前，只有唐凯等（Tang et al.）利用一个参数型谢泼德（Shephard）投入距离函数对澳大利亚的一个雨养农业区域温室气体的影子价格进行了测算。他们认为，雨养农业温室气体减排在成本上具有一定的优势。④ 目前，参数型方向距离函数还未被用于对雨养农业温室气体影子价格的测算中。

本章利用一个允许在技术可行范围内对投入以及厌恶产出同时进行削减的参数型方向性距离函数，对澳大利亚的一个重要雨养农业系统的温室气体减排潜力和影子价格进行测算。本章可能的贡献有两点。首先，对雨养农业系统的温室气体减排的成本有效性进行了综合评估，包括温室气体减排潜力估计和影子价格测算。所测算的结果为设计和实施更加有效的雨养农业温室气体减排政策工具提供了参考。其次，本章构造了一个允许在技术可行范围内对投入以及厌恶产出同时进行削减的参数型方向性距离函数，为生产活动的环境敏感性评价提供了新的思路。投入以及厌恶产出的

---

① Villano, R., Fleming, E., Fleming, P. Evidence of farm-level synergies in mixed-farming systems in the Australian Wheat-Sheep Zone [J]. Agricultural Systems, 2010, 103 (3): 146 - 152.

② Kroon, F. J., Kuhnert, P. M., Henderson, B. L., et al. River loads of suspended solids, nitrogen, phosphorus and herbicides delivered to the Great Barrier Reef lagoon [J]. Marine Pollution Bulletin, 2012, 65 (4 - 9): 167 - 181. Packett, R., Dougall, C., Rohde, K., Noble, R. Agricultural lands are hot-spots for annual runoff polluting the southern Great Barrier Reef lagoon [J]. Marine Pollution Bulletin, 2009, 58 (7): 976 - 986. Packett, R. Rainfall contributes 30% of the dissolved inorganic nitrogen exported from a southern Great Barrier Reef river basin [J]. Marine Pollution Bulletin, 2017, 121 (1 - 2): 16 - 31. Rust, S., Star, M. The cost effectiveness of remediating erosion gullies: a case study in the Fitzroy [J]. Australasian Journal of Environmental Management, 2018, 25 (2): 233 - 247.

③ Srinivasarao, C., Lal, R., Kundu, S., Babu, M. P., Venkateswarlu, B., Singh, A. K. Soil carbon sequestration in rainfed production systems in the semiarid tropics of India [J]. Science of the Total Environment, 2012, 487: 587 - 603.

④ Tang, K., Hailu, A., Kragt, M. E., Ma, C. Marginal abatement costs of greenhouse gas emissions: broadacre farming in the Great Southern Region of Western Australia [J]. Australian Journal of Agricultural and Resource Economics, 2016, 60 (3): 459 - 475.

同时进行削减意味着能够评估雨养农业系统中农业生产活动存在经济—环境双赢的潜力。

4.2 距离函数及其求解方法述评[*]

谢泼德距离函数和方向性产出距离函数是两种被广泛运用于环境敏感性生产分析中的距离函数分析法。[①] 基于卢恩伯格（Luenberger）的效益函数，钱伯斯等（Chambers et al.）引入了方向性距离函数。[②] 谢泼德距离函数假设在存在技术无效的情况下，产出或投入可以沿着方向向量径向增加或减少。[③] 谢泼德（1970）的投入距离函数或者产出距离函数可以在产出水平既定的情况下测量投入的最大缩减程度或者在投入水平既定的情况下测量产出的最大扩张程度。[④] 钱伯斯等（1996）提出的方向性距离函数更具有一般性，它可以测量具有多投入、多产出（期望产出或者厌恶产出或者两者同时存在）生

---

[*] 此节部分内容引自王彪（2012）和魏方庆（2018）。王彪：《中国地方政府财政支出效率研究》，华中科技大学 2012 年博士学位论文。魏方庆：《基于非径向距离函数 DEA 模型的效率评价方法研究》，中国科学技术大学 2018 年博士学位论文。

[①] Tang, K., Hailu, A., Kragt, M. E., Ma, C. Marginal abatement costs of greenhouse gas emissions: broadacre farming in the Great Southern Region of Western Australia [J]. Australian Journal of Agricultural and Resource Economics, 2016, 60 (3): 459 – 475. Tang, K., Gong, C., Wang, D. Reduction potential, shadow prices, and pollution costs of agricultural pollutants in China [J]. Science of the Total Environment, 2016, 541: 42 – 50. Molinos-Senante, M., Sala-Garrido, R. How much should customers be compensated for interruptions in the drinking water supply? [J]. Science of The Total Environment, 2017, 586: 642 – 649. Liu, J. Y., Feng, C. Marginal abatement costs of carbon dioxide emissions and its influencing factors: A global perspective [J]. Journal of Cleaner Production, 2018, 170: 1433 – 1450.

[②] Luenberger, D. G. Benefit functions and duality [J]. Journal of Mathematical Economics, 2004, 21 (5): 461 – 481. Chambers R. G., Chung, Y., Färe, R. Benefit and distance functions [J]. Journal of Economic Theory, 1996, 70 (2): 407 – 419.

[③] Mosavi, S. H., Alipour, A., Shahvari, N. Liberalizing energy price and abatement cost of emissions: Evidence from Iranian agro-environment [J]. Journal of Agricultural Science and Technology, 2017, 19 (3): 511 – 523. Maziotis, A., Villegas, A., Molinos-Senante, M. The cost of reducing unplanned water supply interruptions: A parametric shadow price approach [J]. Science of the Total Environment, 2020, 719, 137487.

[④] Shephard, R. W. Theory of Cost and Production Functions [M]. Princeton: Princeton University Press, 1970.

产单元的效率和生产力。① 不同于谢泼德距离函数，方向距离函数在效率测量时可以同时考虑期望产出的扩张和投入（或者非期望产出）的缩减。② 方向性产出距离函数是最为广泛运用的一种方向性距离函数，其假设期望产出的增加和厌恶产出的减少能够同时实现，是短缺函数的一个变种。③ 尽管这些距离函数能够较好地处理传统的生产分析问题，但它们却无法处理同时减少投入以及厌恶产出的情况。为了弥补这一不足，本章使用了一个允许在技术可行范围内对投入以及厌恶产出同时进行削减的方向性距离函数。这为生产活动的环境敏感性评价提供了新的思路。

在基于方向性距离函数的效率评价模型和应用中，一个非常重要但至今还没有得到完美解决的问题就是如何选择合适的方向对决策单元进行效率评价。④ 已有文献证实方向向量的选择会影响决策单元的技术效率、规模效率、生产率变化⑤以及厌恶产出的影子价格⑥。方向性距离函数的研究主要分为两大类：一种是决策者提前选择方向，称为外生方向方向性距离函数模型；另一种则是通过某种内在的机制（如成本最小化、利润最大化等）选择方向，称为内生方向方向性距离函数模型。不同的方向选择方法会提供给决策者不同的效率和生产力评价依据，这主要取决于研究者的目的以及技术发展水平。⑦

外生方向方向性距离函数模型主要包括任意方向方向性距离函数模型以及条件方向方向性距离函数模型两种。目前在方向性距离函数任意方向选择

---

① Chambers R. G., Chung, Y., Färe, R. Benefit and distance functions [J]. Journal of Economic Theory, 1996, 70 (2): 407 – 419.

② Wu, D., Li, S., Liu, L., Lin, J., Zhang, S. Dynamics of pollutants' shadow price and its driving forces: An analysis on China's two major pollutants at provincial level [J]. Journal of Cleaner Production. DOI: 10.1016/j.scitotenv.2020.136690.

③ Färe, R., Grosskopf, S., Weber, W. L. Shadow prices and pollution costs in US agriculture [J]. Ecological Economics, 2006, 56 (1): 89 – 103.

④⑦ Wang, K., Xian, Y., Lee, C. Y., Wei, Y. M., Huang, Z. On selecting directions for directional distance functions in a non-parametric framework: a review [J]. Annals of Operations Research, 2019, 278: 43 – 76.

⑤ Vardanyan, M., Noh, D. Approximating pollution abatement costs via alternative specifications of a multi-output production technology: A case of the US electric utility industry [J]. Journal of Environmental Management, 2006, 80 (2): 177 – 190. Agee, M. D., Atkinson, S. E., Crocker, T. D. Child maturation, time-invariant, and time-varying inputs: their interaction in the production of child human capital [J]. Journal of Productivity Analysis, 2012, 38 (1): 29 – 44.

⑥ Leleu, H. Shadow pricing of undesirable outputs in nonparametric analysis [J]. European Journal of Operational Research, 2013, 231 (2): 474 – 480.

方法中，普遍使用的两类方向包括：(1) 以被评价决策单元的投入/产出向量作为方向向量，即 $(-g_x, g_y) = (-x_k, y_k)$，$k = 1, 2, \cdots, n$。① 在方向向量为投入/产出向量的情况下，每个决策单元都有自己特定的方向向量，所有决策单元基于各自特定的方向进行效率评价。② (2) 以 $(1, 1, \cdots, 1)_{m+s}$ 作为方向向量。③ 在这种情况下，所有的决策单元具有相同的方向向量，并基于这同一方向向量进行效率评价。④ 以上两种方向性距离函数模型不需要对方向向量进行任何特定的假设，但是存在一些缺点：首先，以上两种方向向量的选择没有特定的经济含义、政策意义以及理论依据；其次，在投入冗余或者产出不足的情况下，方向向量的任意选择会高估被评价决策单元的效率；再次，以投入/产出作为方向向量，所有决策单元的评价标准不统一；最后，如果以 $(1, 1, \cdots, 1)_{m+s}$ 作为方向向量对决策单元进行效率评价的话，效率值不具备单位不变性。⑤

通过设定具体的评价情境（如影子价格的测量）使所有被评价决策单元具有可比性、政策意义等，基于某种条件选择方向的方向性距离函数模型是对任意方向方向性距离函数模型的进一步扩展。尼等（Lee et al.）运用环境

---

① Chambers R. G., Chung, Y., Färe, R. Benefit and distance functions [J]. Journal of Economic Theory, 1996, 70 (2): 407 – 419.

② Kumar, S. Environmentally sensitive productivity growth: A global analysis using Malmquist-Luenberger index [J]. Ecological Economics, 2006, 56 (2): 280 – 293. Oum, T. H., Pathomsiri, S., Yoshida, Y. Limitations of DEA – based approach and alternative methods in the measurement and comparison of social efficiency across firms in different transport modes: An empirical study in Japan [J]. Transportation Research Part E: Logistics and Transportation Review, 2013, 57: 16 – 26. Hampf, B., Krüger, J. J. Technical efficiency of automobiles-A nonparametric approach incorporating carbon dioxide emissions [J]. Transportation Research Part D: Transport & Environment, 2014, 33: 47 – 62.

③ Färe, R., Grosskopf, S., Weber, W. L. Shadow prices and pollution costs in US agriculture [J]. Ecological Economics, 2006, 56 (1): 89 – 103.

④ Färe, R., Grosskopf, S., Noh, D. W., Weber, W. Characteristics of a polluting technology: Theory and practice [J]. Journal of Econometrics, 2005, 126 (2): 469 – 492. Picazo-Tadeo, A. J., Reig-Martínez, E., Hernández-Sancho, F. Directional distance functions and environmental regulation [J]. Resource & Energy Economics, 2005, 27 (2): 131 – 142. Bellenger, M. J., Herlihy, A. T. An economic approach to environmental indices [J]. Ecological Economics, 2009, 68: 2216 – 2223. Halkos, G. E., Tzeremes, N. G. A conditional directional distance function approach for measuring regional environmental efficiency: Evidence from UK regions [J]. European Journal of Operational Research, 2013, 227 (1): 182 – 189.

⑤ Wang, K., Xian, Y., Lee, C. Y., Wei, Y. M., Huang, Z. On selecting directions for directional distance functions in a non-parametric framework: A review [J]. Annals of Operations Research, 2019, 278: 43 – 76.

以及产品的年计划作为方向向量,目的在于测量生产过程中污染物的影子价格,同时考虑了污染物的处理计划。① 德沃等(Dervaux et al.)和西玛等(Simar et al.)分别采用所有方向性距离函数的投入/产出的平均值作为方向向量,相较于投入/产出导向模型,该模型增强了所有决策单元的可比性。② 基于政策意义目的的方向性距离函数模型也得到了广泛的研究,例如,纽基和布拉沃尤维达(Njuki and Bravo-Ureta)以环境管理为准则,选择(1, 1, …, 1)$_{m+s}$为方向向量同时实现期望产出和厌恶产出的增减。③ 尽管基于条件选择的方向性距离函数模型避免了任意方向方向性距离函数模型中方向向量选取的随意性,但是条件方向方向性距离函数模型仍然缺少一些经济含义或者理论基础。④

任意方向方向性距离函数模型和条件方向方向性距离函数模型需要决策者根据自己的研究目的提前假定方向向量,因而评价结果的客观性和合理性会受到一定的影响。基于理论优化的方向性距离函数模型和市场导向的方向性距离函数模型应运而生。

理论优化方向性距离函数模型通常是寻找无效决策单元到前沿面的一个特定投影点。与之前方向性距离函数模型不同的是,理论优化方向性距离函数模型是基于某种理论基础选择方向向量。正是因为有理论上的支持,这类方向性距离函数模型比前述提到的任意方向以及条件方向方向性距离函数模型更为合理。基于此,福莱和哈克(Frei and Harker)首次提出最小范数模型,用于测量无效决策单元到支撑性超平面(非前沿面)的距离,如果该超

---

① Lee, J. D., Park, J. B., Kim, T. Y. Estimation of the shadow prices of pollutants with production/environment inefficiency taken into account: a nonparametric directional distance function approach [J]. Journal of Environmental Management, 2002, 64 (4): 365–375.

② Dervaux, B., Leleu, H., Minvielle, E., Valdmanis, V., Aegerter, P., Guidet, B. Performance of French intensive care units: A directional distance function approach at the patient level [J]. International Journal of Production Economics, 2009, 120 (2): 585–594. Simar, L., Vanhems, A., Wilson, P. W. Statistical inference for DEA estimators of directional distances [J]. European Journal of Operational Research, 2012, 220 (3): 853–864.

③ Njuki, E., Bravo-Ureta, B. E. The economic costs of environmental regulation in U. S. dairy farming: A directional distance function approach [J]. American Journal of Agricultural Economics, 2015, 97 (4): 1087–1106.

④ Wang, K., Xian, Y., Lee, C. Y., Wei, Y. M., Huang, Z. On selecting directions for directional distance functions in a non-parametric framework: A review [J]. Annals of Operations Research, 2019, 278: 43–76.

平面与前沿面是分开的，那么该模型就无法提供一个准确的效率值。① 随后，贝克和尼（Baek and Lee）提出最短距离模型，用于测量无效决策单元到前沿面的最短距离，在前沿面上找到与被评价决策单元最接近的目标值。② 与之相反，费尔和古斯考夫（Färe and Grosskopf）以及阿德勒和沃尔塔（Adler and Volta）分别提出加性方向性距离函数模型来度量无效决策单元到前沿面的最远距离。③ 相较于最近距离函数模型，最远距离函数模型可以帮助无效决策单元找到最大的改进空间，因而具有更高的可信度。然而，即便无效决策单元可以找到自己的最大可能改进潜能，由于受到实际生产水平的限制，无效决策单元并不能实现这一改进。

市场导向方向性距离函数模型综合考虑了理论优化方向以及一些特殊的经济含义，如成本最小化、利润最大化以及边际成本最大化等，需要注意的是该模型更多的是选择具有经济含义的方向。④ 以成本最小化为例，瑞和穆克吉（Ray and Mukherjee）构建了成本前沿面，将被评价决策单元的投入成本信息考虑在内，寻找使成本最小的方向。⑤ 成本最小化方向性距离函数模型得到的效率值可以帮助企业制定生产计划以及进行投资决策。考虑到每个决策单元希望实现利润最大化这一目标，佐菲尔等（Zofio et al.）提出了基于利润效率的方向性距离函数模型，该模型可以帮助无效决策单元投影到前沿面最近的利润最大化目标值，但是利润最大化方向性距离函数模型得到的目标值是随着时间变化的，因此，该模型得到的目标值是不固定的。⑥ 在投入成本以及产出价格信息确定的情况下，尼提出了边际生产力方向性距离函

---

① Frei, F. X., Harler, P. T. Projections onto efficient frontiers: Theoretical and computational extensions to DEA [J]. Journal of Productivity Analysis, 1999, 11 (3): 275 – 300.

② Baek, C., Lee, J. D. The relevance of DEA benchmarking information and the Least-Distance Measure [J]. Mathematical & Computer Modelling, 2009, 49: 265 – 275.

③ Adler, N., Volta, N. Accounting for externalities and disposability: A directional economic environmental distance function [J]. European Journal of Operational Research, 2016, 250 (1): 314 – 327. Färe, R., Grosskopf, S. Directional distance functions and slacks-based measures of efficiency: Some clarifications [J]. European Journal of Operational Research, 2010, 206 (3): 702 – 702.

④ Wang, K., Xian, Y., Lee, C. Y., Wei, Y. M., Huang, Z. On selecting directions for directional distance functions in a non-parametric framework: A review [J]. Annals of Operations Research, 2019, 278: 43 – 76.

⑤ Ray, S. C., Mukherjee, K. Decomposition of cost competitiveness in US manufacturing: Some state-by-state comparisons [J]. Indian Economic Review, 2000, 35 (2): 133 – 153.

⑥ Zofio, J. L., Pastor, J. T., Aparicio, J. The directional profit efficiency measure: On why profit inefficiency is either technical or allocative [J]. Working Papers in Economic Theory, 2010, 40 (3): 257 – 266.

数模型，该模型基于边际利润最大化方向对决策单元进行评价。① 概括来讲，成本最小化方向性距离函数模型和边际利润最大化方向性距离函数模型中的成本、价格信息是外生信息，但是利润最大化方向性距离函数模型中的价格信息可以是内生信息，也可以是外生信息，这主要取决于不同的市场情境。

除了一般均衡分析法、优化法以及成本效益分析法，已有研究多使用非参数估计法和参数估计法来求解距离函数。数据包络分析（data envelopment analysis, DEA）是最常见的非参数估计法。该方法直接基于一组特定决策单位的数据而不是某种特定的函数形式来界定生产可能性边界，某个决策单位的实际生产点与最优生产边界的距离即反映了这一决策单位的无效率。DEA 通过求解线性规划找出一个包络所有实际生产点的最小凸锥，因此，其生产可能性边界的形式是由实际投入与产出情况下的最优生产点集合组成的折线。作为非参数方法的典型代表，DEA 是评价一组具有多投入、多产出同质生产单元相对效率的线性规划方法。相较于其他效率评价方法，DEA 不需要提前确定投入、产出之间的函数关系，不需要对生产函数形式进行提前假设，更不用去估计投入、产出的权重参数，因而规避了效率评价结果的主观影响。此外，DEA 对生产单元进行绩效测量时，还能够对所有生产单元进行排序，帮助无效生产单元确定标杆和改进方向，且对数据形式与质量没有过于严格的要求。

由于 DEA 在评价多投入、多产出决策单元相对效率方面具有独特优势，自 1978 年提出之后，经典的 DEA 模型（如 CCR② 和 BCC③）及其改进模型（如 Additive Model④，Cross-efficiency⑤，Super-efficiency⑥，SBM⑦ 和

---

① Lee, C. Y. Meta-data envelopment analysis: Finding a direction towards marginal profit maximization [J]. European Journal of Operational Research, 2014, 237 (1): 207 – 216.

② Charnes, A., Cooper, W. W., Rhodes, E. Measuring the efficiency of decision making units [J]. European journal of operational research, 1978, 2 (6): 429 – 444.

③ Banker, R. D., Charnes, A., Cooper, W. W. Some models for estimating technical and scale inefficiencies in data envelopment analysis [J]. Management Science, 1984, 30 (9): 1078 – 1092.

④ Charnes, A., Cooper, W. W., Golany, B., Seiford, L., Stutz, J. Foundations of data envelopment analysis for Pareto-Koopmans efficient empirical production functions [J]. Journal of Econometrics, 1985, 30: 91 – 107.

⑤ Sexton, T. R., Silkman, R. H., Hogan, A. J. Data envelopment analysis: Critique and extensions [J]. New Directions for Evaluation, 2010, 1986 (32): 73 – 105.

⑥ Andersen, P., Petersen, N. C. A procedure for ranking efficient units in data envelopment analysis [J]. Management Science, 1993, 39 (10): 1261 – 1264.

⑦ Tone, K. A slacks-based measure of efficiency in data envelopment analysis [J]. European Journal of Operational Research, 2001, 130 (3): 498 – 509.

Network DEA①）已经被广泛应用于不同领域，如教育、银行、投资、医疗、体育、制造业、物流、通信②等。由于 DEA 在理论方法上得到不断创新和拓展，其可以解决的实际问题越来越多，已经逐渐成为管理科学、系统工程以及经济学中一个非常重要的分析工具，被大量运用于涉及生产效率评价的相关研究中。③

然而，DEA 的一个重大缺陷是假设不存在随机误差的影响，由于忽略潜在的偏误，随机误差可能会包括效率项的估计中，特别是如果处于效率边界上的决策单位存在随机误差，就会影响所有决策单位的效率估计。④ 同时，DEA 运用存在多个决策单元效率值为 1 的情况，无法进行决策单元之间的具体比较。而且，DEA 不能方便地检验结果的显著性，所求出的生产函数是不

---

① Kao, C. Network data envelopment analysis: A review [J]. European Journal of Operational Research, 2014, 239 (1): 1 – 16.

② Grosskopf, S., Hayes, K. J., Taylor, L. L., Weber, W. L. Anticipating the consequences of school reform: A new use of DEA [J]. Management Science, 1999, 45 (4): 608 – 620. Seiford, L. M., Zhu, J. Profitability and marketability of the top 55 US commercial banks [J]. Management Science, 1999, 45 (9): 1270 – 1288. Lamb J. D., Tee, K. H. Data envelopment analysis models of investment funds [J]. European Journal of Operational Research, 2012, 216 (3): 687 – 696. Banker, R. D., Conrad, R. F., Strauss, R. P. A comparative application of data envelopment analysis and translog methods: an illustrative study of hospital production [J]. Management Science, 32 (1): 30 – 44. Yang, F., Wu, D. D., Liang, L., O'Neill, L. Competition strategy and efficiency evaluation for decision making units with fixed-sum outputs [J]. European Journal of Operational Research, 2011, 212 (3): 560 – 569. Wahab, M. I. M., Wu, D., Lee, C. G. A generic approach to measuring the machine flexibility of manufacturing systems [J]. European Journal of Operational Research, 2008, 186 (1): 137 – 149. Xu, J., Li, B., Wu, D. Rough data envelopment analysis and its application to supply chain performance evaluation [J]. International Journal of Production Economics, 2009, 122 (2): 628 – 638. Cooper, W. W., Park, K. S., Yu, G. An illustrative application of IDEA (imprecise data envelopment analysis) to a Korean mobile telecommunication company [J]. Operations Research, 2001, 49 (6): 807 – 820.

③ 杨国梁：《DEA 模型与规模收益研究综述》，载《中国管理科学》2015 年第 S1 期。Färe, R., Grosskopf, S., Pasurka, C. A. Environmental production functions and environmental directional distance functions [J]. Energy, 2007, 32 (7): 1055 – 1066. Wang, S., Chu, C., Chen, G., Peng, Z., Li, F. Efficiency and reduction cost of carbon emissions in China: A non-radial directional distance function method [J]. Journal of Cleaner Production, 2016, 113: 624 – 634. Emrouznejad, A., Yang, G. L. A survey and analysis of the first 40 years of scholarly literature in DEA: 1978 – 2016 [J]. Socio-Economic Planning Sciences, 2018, 61: 4 – 8. 于斌斌：《产业结构调整如何提高地区能源效率？——基于幅度与质量双维度的实证考察》，载《财经研究》2017 年第 1 期。韩兵、苏屹、李彤等：《基于两阶段 DEA 的高技术企业技术创新绩效研究》，载《科研管理》2018 年第 3 期。王赵琛：《24 所部属高校科技成果转化效率的 DEA 分析》，载《科研管理》2020 年第 4 期。

④ 郝睿：《经济效率与地区平等：中国省际经济增长与差距的实证分析（1978 – 2003）》，载《世界经济文汇》2006 年第 2 期。

可微的。此外，DEA对数据的敏感性较大，一个异常数据的出现容易引起效率前沿面的改变，进而导致结果的较大变化。

参数估计法是事先假定具体的函数形式，以样本为基础，利用多元统计分析方法，对事先假定的函数参数进行求解的一种经济计量方法。参数估计法根据在假定具体的函数形式时对有关误差项的不同假设，又可以分为确定性方法和随机性方法。确定性方法假设函数的误差项全部来自人为的可控因素，属于技术无效率，并且无效率决策单元和有效率决策单元之间的差额值服从单一分布函数。与确定性方法假设不同，随机性方法假设误差项不仅来自人为可控的技术无效率误差，还来自无法人为控制的随机性误差。技术无效率误差反映的是决策单元的无效率程度，它服从单边概率分布；而随机性误差服从对称性的双边概率分布。参数方法中的随机性方法根据对前沿函数中无效率项分布的不同假设，又可以分为随机前沿分析方法、厚边界分析法和自由分布法。[①]

随机前沿分析方法认为，主要是因为技术无效率项和随机误差项的存在导致了部分决策单元与效率前沿决策单元发生了偏离，效率前沿决策单元是指在给定的技术条件、投入要素价格和外部环境下，能以一定成本获得最大产出（或以最小成本获得一定产出）的决策单元，它们是从产出最大化（或成本最小化）角度构造出的一组处于最佳状态的决策单元，并且假定效率前沿决策单元的效率值为，而样本中其他决策单元的效率值都小于1。厚边界分析法与随机前沿分析方法相比，其最主要的特征就是不事先假设无效率项和随机误差项的概率分布，取而代之的是将所有样本决策单元按效率值排序，然后把排序后的决策单元平均分为四组，把效率最好和效率最差的两组样本作为研究重点，并且进一步假定这两组样本决策单元的组内差异是因为随机误差因素所造成的，而组间差异则是由无效率因素引起的。[②] 自由分布法认为，主要是因为低效率项和随机误差项的存在导致了部分决策单元与效率前沿决策单元发生了偏离，因而还要在函数中引入低效率项和随机误差项，该方法通过极大似然估计方法得到生产函数或成本函数所有参数的估计值，最后计算出样本中各决策单元的低效率值。低效率值最小的决策单元即为样本中的最佳决策单元，也就是效率前沿决策单元，并假设其相对效率值为1。然后，将效率前沿决策单元的低效率值与样本中其他决策单元的低效率值相

[①][②] 王彪：《中国地方政府财政支出效率研究》，华中科技大学2012年博士学位论文。

比,得到其他决策单元的相对效率值。自由分布法的前提假设是所有决策单元的生产经营效率在研究期间内是稳定的,并且没有指定低效率项和随机误差项的概率分布形态。

与 DEA 不同的是,参数估计法需要先明确生产函数的形式,然后再利用数学规划法求解有关参数。由于参数估计法具有便于设定相关理论约束以及允许随机误差存在的优势,参数估计法也开始被广泛地运用到不同问题的分析中。① 陈诗一②在对国内工业二氧化碳影子价格的测算中,对比了 DEA 与参数估计法之间的区别和联系,其实证分析结果与尼等和费尔等类似③,两种方法度量结果的大小和变化趋势相近,生产率指数非常相似,但参数化方法估计值更为可靠。

4.3 研究方法与数据

4.3.1 参数型方向性距离函数及影子价格的推导

在雨养农业生产中,农民生产农产品(期望产出)需要不同的投入,如农地、劳动以及资本等。这一生产过程往往也会产生厌恶产出,如排放的温室气体。假设 x 为农业投入向量,y 为期望产出向量,b 为厌恶产出向量(温

---

① Coggins, J. S., Swinton, J. R. The price of pollution: a dual approach to valuing SO_2 allowances [J]. Journal of Environmental Economics and Management, 30 (1): 58 – 72. Tang, K., Hailu, A., Kragt, M. E., Ma, C. Marginal abatement costs of greenhouse gas emissions: broadacre farming in the Great Southern Region of Western Australia [J]. Australian Journal of Agricultural and Resource Economics, 2016, 60 (3): 459 – 475. Tang, K., Gong, C., Wang, D. Reduction potential, shadow prices, and pollution costs of agricultural pollutants in China [J]. Science of the Total Environment, 2016, 541: 42 – 50. Molinos-Senante, M., Sala-Garrido, R. How much should customers be compensated for interruptions in the drinking water supply? [J]. Science of The Total Environment, 2017, 586: 642 – 649.

② 陈诗一:《工业二氧化碳的影子价格:参数化和非参数化方法》,载《世界经济》2010 年第 8 期。

③ Lee, J. D., Park, J. B., Kim, T. Y. Estimation of the shadow prices of pollutants with production/environment inefficiency taken into account: a nonparametric directional distance function approach [J]. Journal of Environmental Management, 2002, 64 (4): 365 – 375. Färe, R., Grosskopf, S., Noh, D. W., Weber, W. Characteristics of a polluting technology: Theory and practice [J]. Journal of Econometrics, 2005, 126 (2): 469 – 492.

室气体），则农业生产集可定义为：

$$T(x) = \{(x,y,b): x\text{ 可生产出}(y,b)\} \qquad (4-1)$$

其中，$T(x)$ 是一个有界闭集。$T(x)$ 满足以下几个性质。[①]

(a) 凸性[②]。

(b) 投入和期望产出的强可处置性[③]，即：如果 $(x,y,b) \in T, (x', -y') \geq (x, -y)$，那么 $(x', y', b) \in T$。

(c) 厌恶产出的弱可处置性[④]，即：如果 $(x, y, b) \in T$ 且 $0 \leq \eta \leq 1$，则 $(x, \eta y, \eta b) \in T$。具体来说，期望产出和厌恶产出按相同比例缩减之后还在农业生产集之内。换句话讲，厌恶产出的减少必然伴随着期望产出的减少或者是投入的增加，即减少厌恶产出需要付出一定的成本[⑤]。

(d) 期望产出和厌恶产出的零结合性（Null-Jointness），即：如果 $(x, y, b) \in T$ 且 $b = 0$，则 $y = 0$。具体而言，厌恶产出不存在时，期望产出也不存在，也可以说，在得到期望产出的同时必然会得到厌恶产出，或者说，厌恶产出是农业生产进行的必然产物。

代表 $T(x)$ 函数特征的方向性距离函数可定义为：

$$\vec{D}_T(x,y,b;g) = \max\{\beta: (x - \beta g_x, y + \beta g_y, b - \beta g_b) \in T, \beta \in R_+\} \qquad (4-2)$$

其中，$g = (g_x, g_y, g_b) \in R_+^M \times R_+^N \times R_+^J$ 表示的是方向向量，即 (x, y, b) 沿着方向 g 通过同时增加期望产出而削减投入以及厌恶产出最终将到达技术边界 T（生产前沿）。若这一改变是技术可行的，则函数值的范围为 $[0, \infty)$。对于 $\vec{D}_T$ 的有关性质，钱伯斯等（Chambers et al.）、海努和维曼（Hailu and Veeman）以及费尔等有详细描述。[⑥]

[①][③][④] Färe, R., Grosskopf, S., Weber, W. L. Shadow prices and pollution costs in US agriculture [J]. Ecological Economics, 2006, 56 (1): 89 – 103.

[②] Shephard, R. W. Theory of Cost and Production Functions [M]. Princeton: Princeton University Press, 1970.

[⑤] Färe, R., Grosskopf, S., Noh, D. W., Weber, W. Characteristics of a polluting technology: Theory and practice [J]. Journal of Econometrics, 2005, 126 (2): 469 – 492.

[⑥] Chambers, R. G., Chung, Y., Färe, R. Profit, directional distance functions, and Nerlovian efficiency [J]. Journal of Optimization Theory and Applications, 1998, 98 (2): 351 – 364. Hailu, A., Veeman, T. S. Non-parametric productivity analysis with undesirable outputs: an application to the Canadian pulp and paper industry [J]. American Journal of Agricultural Economics, 2001, 83 (3): 605 – 616. Färe, R., Grosskopf, S., Weber, W. L. Shadow prices and pollution costs in US agriculture [J]. Ecological Economics, 2006, 56 (1): 89 – 103.

现在我们考虑$\vec{D}_T$的一个特殊形式，即g_y为0，那么$\vec{D}_T$可以被改写为：

$$\vec{D}_D(x,y,b;g) = \max\{\beta:(x-\beta g_x, b-\beta g_b) \in T, \beta \in R_+\} \quad (4-3)$$

$\vec{D}_D$反映了同时减少投入与厌恶产出直到现有技术水平的程度。$\vec{D}_D$以及方向性产出距离函数都是方向性距离函数$\vec{D}_T$的特例。

厌恶产出的影子价格可以从利润最大化函数中求得：

$$\pi(x,y,b,p,q,r) = \max\{qy - px - rb : \vec{D}_T(x,y,b;-g_x,g_y,-g_b) \geq 0\}$$
$$(4-4)$$

其中，$p = (p_1,\cdots,p_m) \in R_+^M$、$q = (q_1,\cdots,q_n) \in R_+^N$和$r = (r_1,\cdots,r_j) \in R_+^J$分别为期望产出、投入和厌恶产出的价格。$\lambda = qg_y - pg_x - rg_b$为式（4-4）的拉格朗日乘子。假定$g_y = 0$即没有考虑期望产出$y$的变化。给定第$n$种期望产出的价格$q_n$，则第$j$种厌恶产出的影子价格$r_j$可表示为：

$$r_j = -q_n \left(\frac{\partial \vec{D}_D(x,y,b;-g_x,-g_b)/\partial b_j}{\partial \vec{D}_D(x,y,b;-g_x,-g_b)/\partial y_m} \right), \quad j = 1,\cdots,J \quad (4-5)$$

影子价格是指在生产消耗、产品价格等已知条件固定的情况下，对资源合理配置和优化组合后，某种资源增加一单位所能带来的边际收益。作为一种非期望产出，厌恶产出的影子价格表示每减少一单位厌恶产出排放导致的收益或产出的减少量。影子价格由于反映了期望产出与厌恶产出之间的权衡，因而可以被认为是厌恶产出的机会成本或边际减排成本，可用于估计边际减排成本。当样本数据越微观时，二氧化碳的影子价格越接近真实的边际减排成本。① 此外，在方向性距离函数值与影子价格之间存在负向关系。当雨养农业生产者的生产效率越高时（即方向性距离函数值越低），减少额外一单位厌恶产出的难度要高于那些生产效率较低的生产者。这意味着，在雨养农业中，与生产效率较低的生产者相比，生产效率较高的生产者不得不舍弃更多的期望产出以换取一单位厌恶产出的减少。因此，对生产效率较高的生产者而言，更高的机会成本将体现为更高的厌恶产出影子价格。

---

① 袁鹏、程施：《我国工业污染物的影子价格估计》，载《统计研究》2011年第9期。周鹏、周迅、周德群：《二氧化碳减排成本研究述评》，载《管理评论》2014年第11期。

4.3.2 实证说明

本章使用一般化的二次型形式来对方向性距离函数$\vec{D}_D$进行参数化。[①] 利用方向向量$g=(1,1)$简化参数化过程。$g=(1,1)$描述的是对投入与厌恶产出进行单位削减。假设在$t=1,\cdots,T$时期内有$k=1,\cdots,K$名生产者，则二次型方向性距离函数$\vec{D}_D$可表示为：

$$\vec{D}_D^t(x_k^t, y_k^t, b_k^t; 1,1) = \alpha_0 + \sum_{m=1}^{M} \alpha_m x_{mk}^t + \sum_{n=1}^{N} \beta_n y_{nk}^t + \sum_{j=1}^{J} \gamma_j b_{jk}^t$$
$$+ \frac{1}{2}\sum_{m=1}^{M}\sum_{m'=1}^{M} \alpha_{mm'} x_m x_{m'} + \frac{1}{2}\sum_{n=1}^{N}\sum_{n'=1}^{N} \beta_{nn'} y_{nk}^t y_{n'k}^t$$
$$+ \frac{1}{2}\sum_{j=1}^{J}\sum_{j'=1}^{J} \gamma_{jj'} b_j b_{j'} + \sum_{m=1}^{M}\sum_{n=1}^{N} \delta_{mn} x_{mk}^t y_{nk}^t$$
$$+ \sum_{m=1}^{M}\sum_{j=1}^{J} \tau_{mj} x_{mk}^t b_{jk}^t + \sum_{n=1}^{N}\sum_{j=1}^{J} \varphi_{nj} y_{nk}^t b_{jk}^t \quad (4-6)$$

其中，$\alpha_{mm'} = \alpha_{m'm}$，$m \neq m'$，$\beta_{nn'} = \beta_{n'n}$，$n \neq n'$，且$\gamma_{jj'} \neq \gamma_{j'j}$，$j \neq j'$。

$\vec{D}_D$各参数的估计通过估计以下数学规划问题来实现：

$$\min \sum_{t=1}^{T}\sum_{k=1}^{K} [\vec{D}_D^t(x_k^t, y_k^t, b_k^t; 1,1) - 0] \quad (4-7)$$

约束条件包括单调性、转移性以及技术可行性。详细约束条件的内容参见海努和钱伯斯（Hailu and Chambers）。[②] 为了消除收敛性问题，将投入与产出的数据用其各自的均值进行标准化。[③] 这样的处理意味着一个假设的雨养农业农场使用平均投入来生产平均产出。

二次型方向性距离函数$\vec{D}_D$参数的求解使用 R 软件中的 APEAR 程序包。[④]

[①] Färe, R., Grosskopf, S., Weber, W. L. Shadow prices and pollution costs in US agriculture [J]. Ecological Economics, 2006, 56 (1)：89-103.

[②] Hailu, A., Chambers, R. G. A Luenberger soil-quality indicator [J]. Journal of Productivity Analysis, 2012, 38 (2)：145-154.

[③] Färe, R., Grosskopf, S., Pasurka Jr, C. A. Environmental production functions and environmental directional distance functions [J]. Energy, 2007, 32 (7)：1055-1066.

[④] Hailu, A. APEAR：A package for productivity and efficiency analysis in R (version0.1) [Z]. UWA, 2013. Available from URL：http://ahailu.are.uwa.edu.au.

方向性距离函数$\overrightarrow{D}_D$的函数值为体现技术效率水平的、投入与产出能够同时减少的最大程度。若函数值为0，则表示生产是完全有效的；函数值为正则表示存在技术无效；函数值越小，则表示技术无效的程度越低。

雨养农业温室气体的技术减排潜力可用式（4-8）进行估算：

$$\Delta b_{kt} = b_{kt} - (b_{kt} - \beta_{kt} g_b) \tag{4-8}$$

其中，b_{kt}、β_{kt}和g_b分别为温室气体排放量、农场k在t时期式（4-6）的估计值和温室气体排放的方向向量。Δb_{kt}描述的是当雨养农业生产完全有效时，农场k在t时期所能实现的温室气体排放量的最大减少值。本章使用所估计的农场温室气体技术减排潜力与实际排放量之间的比值，以方便在不同的农场以及年份间进行比较。这样可以减少由样本农场温室气体排放量以及生产规模上存在的差异所引起的负面影响。

所测算的减排潜力与影子价格之间存在负向关系。式（4-6）与式（4-8）表面技术无效程度越高的农场其β值越大，其减排潜力也越大。对技术无效程度较高的农场而言，其减少额外一单位温室气体排放的难度要低于技术无效程度较低的农场，这也意味着前者的温室气体减排机会成本更低。因此，技术无效程度较高农场的温室气体影子价格由于体现了减排的机会成本而会更低。

4.3.3 数据与变量说明

本章所研究的区域属于澳大利亚西澳大利亚州（Great Southern Region）。该地区是一个典型的雨养农业区。[①] 在该地区，绝大部分农民从事种—畜复合农业经营。该地区是澳大利亚最主要的农业区之一，同时也是西澳大利亚州第二大农产品生产基地。大南区自然环境以及农业生产的相关介绍参见本书第4章。本章分析所使用的样本涵盖该地区的42个家庭农场，时间范围为2006~2013年。样本数据由西澳大利亚州农业与食品部提供。家庭农场投入以及产出数据按照伊斯拉姆等（Islam et al.）的方法进行了处理。农产品（期望产出）包括作物以及畜牧产出。投入包括农地、劳动、资本以及物料

---

① Tang, K., Hailu, A., Kragt, M. E., Ma, C. Marginal abatement costs of greenhouse gas emissions: broadacre farming in the Great Southern Region of Western Australia [J]. Australian Journal of Agricultural and Resource Economics, 2016, 60 (3): 459-475.

与服务。考虑到降水量对雨养农业地区农业生产的重要影响，生长季节降水量也被涵盖到投入中。

本章按照政府间气候变化专门委员会（IPCC）所使用的方法来计算农场温室气体排放量，同时唐凯等（Tang et al.）对部分参数值进行了必要修正，以符合当地雨养农业的自然环境特征。① 具体计算方法及参数值参见本书附录。本章考虑了四类农业温室气体排放源，包括作物秸秆、化肥施用、牲畜（包括肠道发酵和牲畜粪便）以及固氮作物。所有的温室气体排放量参考唐凯等（2016）使用全球暖化潜能值转化为CO_2e，其中，二氧化碳为1，甲烷为21，氧化亚氮为310。②

各家庭农场的农地面积普遍较大，平均为2057公顷，且在研究时期内呈不断扩大的态势，年均增长率超过3%。超过1/2的农地被用作作物生产。研究期间内生长季节降水量的波动较大。各家庭农场的温室气体排放量存在较大差异，为每公顷0.06~1.63tCO_2e。样本农场年均温室气体排放量为809tCO_2e，折合每公顷0.45tCO_2e（见表4-1）。牲畜以及固氮植株（主要是牧草）是最主要的温室气体排放源，其占总排放量的比例分别为63.3%和25.6%。化肥以及作物产生的温室气体分别占总排放量的10.6%和0.5%。以上测算值与现有关于类似雨养农业系统的研究基本一致。③

表4-1　　　　　　　　样本农场投入产出描述性统计

项目	2006年	2007年	2008年	2009年	2010年	2011年	2012年	2013年
劳动（1000澳元）	83	86	89	89	94	111	115	117
农地（公顷）	1832	1886	1972	2045	2079	2159	2224	2256
作物用地（公顷）	1006	1122	1151	1159	1188	1312	1334	1283
畜牧用地（公顷）	826	764	821	887	891	847	889	972
资本（1000澳元）	235	209	220	259	264	376	341	322
物料与服务（1000澳元）	278	298	273	267	344	401	433	399
生长季节降水量（毫米）	407	334	219	279	270	397	237	449

---

①② Tang, K., Hailu, A., Kragt, M. E., Ma, C. Marginal abatement costs of greenhouse gas emissions: broadacre farming in the Great Southern Region of Western Australia [J]. Australian Journal of Agricultural and Resource Economics, 2016, 60: 459-475.

③ Thamo, T., Kingwell, R. S., Pannell, D. J. Measurement of greenhouse gas emissions from agriculture: Economic implications for policy and agricultural producers [J]. Australian Journal of Agricultural and Resource Economics, 2013, 57: 234-252.

续表

项目	2006年	2007年	2008年	2009年	2010年	2011年	2012年	2013年
作物产出（1000澳元）	632	797	441	956	780	1481	903	872
畜牧产出（1000澳元）	221	181	206	169	166	203	220	184
温室气体排放总量（tCO_2e）	791	771	716	720	644	879	894	1053
投入价格指数	100	101	105	110	112	122	123	126
产出价格指数	100	96	96	109	122	123	117	116
作物价格指数	100	92	95	109	123	135	105	96

注：澳元为2013年澳元。价格指数数据来源：澳大利亚农业资源经济科学局（ABARES，2014）。

4.4 研究结果与讨论

本节由四部分组成。首先，给出雨养农业生产的技术无效估计、农场温室气体的减排潜力测算以及利用前述所构造的方向性距离函数法所测算的温室气体排放影子价格；其次，对分析结果进一步讨论。

4.4.1 技术无效水平

样本农场在研究时期内的方向性距离函数$\vec{D}_D$的平均值为0.188。$\vec{D}_D$的年平均值在0.131（2012年）~0.272（2013年）之间波动。由于技术无效水平与$\vec{D}_D$的值之间存在正向关系，上述结果说明在研究时期内技术无效水平没有显著降低。2006年，所有样本农场的平均$\vec{D}_D$值为0.16。这意味着，如果所有家庭农场的生产是完全有效的，农地、劳动、资本以及物料与服务投入分别平均可以减少293公顷、13280澳元、37600澳元以及44480澳元。2013年样本农场的平均$\vec{D}_D$值上升到0.272，意味着在农地、劳动、资本以及物料与服务投入分别存在平均减少614公顷、31824澳元、87584澳元以及108528澳元。以上结果与伊斯拉姆等和唐凯等的研究结果基本一致。[1]

---

[1] Islam, N., Xayavong, V., Kingwell, R. Broadacre farm productivity and profitability in southwestern Australia [J]. Australian Journal of Agricultural and Resource Economics, 2014, 58 (2): 147 - 170. Tang, K., Hailu, A., Kragt, M. E., Ma, C. Marginal abatement costs of greenhouse gas emissions: broadacre farming in the Great Southern Region of Western Australia [J]. Australian Journal of Agricultural and Resource Economics, 2016, 60: 459 - 475.

那么，什么原因导致了技术无效的存在呢？研究期间内最低的技术无效水平出现在2012年，其原因在于对现有生产技术的充分利用和合理的种—畜生产经营结构。虽然该年的生产季节降水量是研究时期内的第二低（237毫米），平均每公顷产量却超过了研究时期内的平均值。这表明，样本农场在相对干旱的年份对现有生产技术运用良好，降低了技术无效水平。此外，农民将其家庭农场种—畜生产经营结构优化到了合理的水平：平均作物种植面积（59.6%）是研究时期内的第二高。合理的种—畜生产经营结构也有助于降低技术无效水平。与此相对的是，较高的温室气体排放量、较高的农业投入价格以及较低的农业产出（见表4-1）使得2013年的技术无效水平达到了研究时期内的最高值。2013年的平均生产季节降水量和每公顷平均温室气体排放量都是研究时期内的最高值。高生产季节降水量促使畜群规模以及畜牧用地（用于种植固氮牧草提供饲料）扩张，因而增加了来自牲畜和牧草的温室气体排放量。高投入、高厌恶产出以及低期望产出意味着该年度的雨养农业生产技术效率相对较低，直接导致了技术无效水平达到研究时期内的最高值。

4.4.2 雨养农业温室气体减排潜力

图4-1描述了样本农场的温室气体减排潜力值。考虑到本章温室气体减排潜力值的测算是基于技术无效水平进行的，温室气体减排潜力值的变动总体上体现了技术无效水平的变化。在研究期间内，样本农场的温室气体减排潜力值多处在10%~40%的范围内。2006~2013年，年均温室气体减排潜力值在14%~33%波动，均值为21%。这说明，如果样本农场的生产是完全有效的，平均可以减少21%即170 tCO_2e 温室气体排放。以上结果与现有一些研究的结果相类似。[1]

2010年的平均减排潜力达到研究时期内的最高点。这很可能是由于没有对现有生产技术进行充分利用而造成的。如前所述，在这个雨养农业区，现有生产技术使用的效率水平对农业生产的技术效率水平有着极大的影响。虽然2010年的生产季节降水量（270毫米）要高于2012年（237毫米），2010

---

[1] Tang, K., Hailu, A., Kragt, M.E., Ma, C. The response of broadacre mixed crop-livestock farmers to agricultural greenhouse gas abatement incentives [J]. Agricultural Systems, 2018, 160: 11-20.

年小麦、大麦以及油菜的单位面积产量却比 2012 年低约 20%。这说明，这一年许多家庭农场的生产没有达到生产技术前沿，排放了大量的不必要的温室气体，因而使得该年的减排潜力值达到研究时期内的最高点。

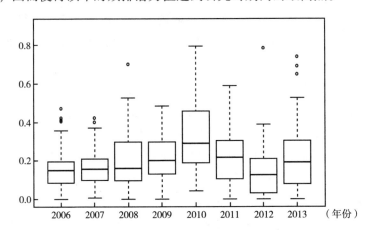

图 4-1　2006~2013 年雨养种—畜复合经营农场的温室气体减排潜力

4.4.3　雨养农业温室气体排放的影子价格

如前所述，影子价格衡量生产中为了减少厌恶产出所舍弃期望产出的机会成本。考虑到样本中所有农场都从事作物生产且作物产值占总产值的比例更大，这里选择反映了作物产出与温室气体排放之间取舍关系的影子价格来代表雨养农业农场温室气体排放的影子价格。

图 4-2 描述了每一年样本农场的影子价格情况。在研究时期内，影子价格大多低于 40 澳元/tCO_2e。2006~2013 年，样本平均影子价格为 17.6 澳元/tCO_2e（约合 88 元人民币/tCO_2e）。年均影子价格在研究时期内呈波动态势，最低值为 7.3 澳元/tCO_2e（2008 年）（约合 36.5 元人民币/tCO_2e），最高值为 27.3 澳元/tCO_2e（2011 年）（约合 136.5 元人民币/tCO_2e）。

引起影子价格波动的一个可能的因素是作物市场价格的变化。2006~2013 年，澳大利亚农场的作物销售价格呈现波动态势（见图 4-3）。变动的作物市场价格意味着为额外减少一单位温室气体排放所舍弃的作物产出的价值也随之变化，导致了雨养农业温室气体减排机会成本的波动。因此，在研究时期内，影子价格的变动趋势与作物市场价格的变动趋势之间存在着一定程度的关联。

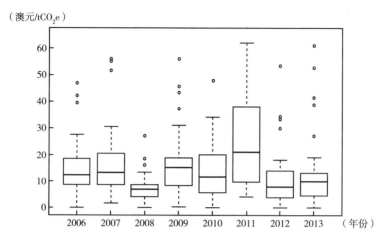

图 4-2 雨养种—畜复合经营农场的温室气体影子价格

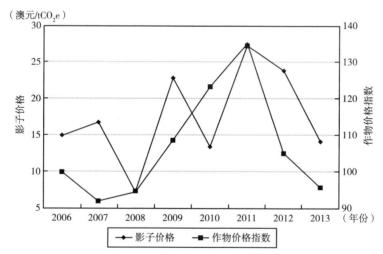

图 4-3 雨养种—畜复合经营农场的平均温室气体影子价格与作物价格指数

注：作物价格指数引自澳大利亚农业资源经济科学局（ABARES, 2014）。

从图 4-4 中可以看出，总体上，平均温室气体影子价格与平均作物产出比重之间存在一个正向关系。从事雨养农业生产的家庭农场的作物产出比重越高，则其温室气体影子价格也越高。对于样本农场而言，作物产出约占总产出的 70%~90%，而温室气体影子价格多在 10~30 澳元/tCO_2e 的范围内。究其原因，主要有以下几点：首先，对于雨养农业区的种—畜复合经营农场而言，高作物产出比重往往意味着低温室气体排放量。作物产出比重较高的家庭农场通常分配更多的农地用于作物种植，因而减少了对畜牧业的农地投入（牧草种植面积）。由于作物与化肥施用是相对次要的温室气体排放源，而牲畜与牧草是最主要的温室气体排放源，其排放强度远高于作物与化肥施用，因此，随着作物种植面积的不断扩大，农场温室气体排放总量会逐步下降。其次，作物产出比重较高的家庭农场与作物产出比重较低的家庭农场相比，前者往往排放更少的温室气体。那么，对于作物产出比重较高的家庭农场而言，额外一单位温室气体占总排放量的比重也越高。这就意味着，为了减少额外一单位的温室气体排放，作物产出比重较高的家庭农场要比作物产出比重较低却排放更多温室气体的家庭农场付出更大的努力，即舍弃更多的期望产出。所以不难看出，作物产出比重较高的家庭农场的影子价格往往更高。

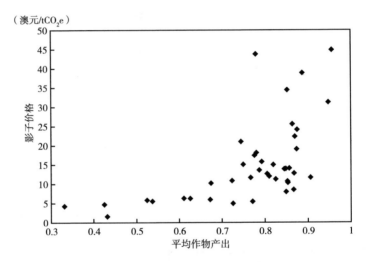

图 4-4　雨养种—畜复合经营农场的平均温室
气体影子价格与平均作物产出比重

现在将所测算的影子价格与已有研究的结果进行比较。唐凯等估计了澳大利亚一个广域农业区域的温室气体影子价格，其平均值为 29.3 澳元/tCO_2e（约合 146.5 元人民币/tCO_2e）。[①] 近期有学者评估了中国农业碳排放影子价格，其平均值为 62.5 元人民币/tCO_2e，约合 12.5 澳元/tCO_2e。[②] 本章所测算

---

① Tang, K., Hailu, A., Kragt, M. E., Ma, C. The response of broadacre mixed crop-livestock farmers to agricultural greenhouse gas abatement incentives [J]. Agricultural Systems, 2018, 160: 11-20.

② Wu, X., Zhang, J., You, L. Marginal abatement cost of agricultural carbon emissions in China: 1993-2015 [J]. China Agricultural Economic Review, 2018, 10 (4): 558-571.

的结果略高于来自中国的结果,但要低于另外一篇澳大利亚文献的结果。除了数据与研究范围上的不同,本章与唐凯等(2016)在研究方法上的差别也可能导致影子价格评价结果的差异。唐凯等(2016)所采用的是参数型谢泼德投入距离函数,其在实质上是本章所采用的方向性距离函数的一个特例。

在澳大利亚现有的排放减量基金框架下,从事雨养农业经营的农民可以进行投标,投标书中明确其自愿采取的碳汇农业措施,以及在采取这些碳汇农业措施后固定的每一吨碳或减少排放的每一吨温室气体所希望得到的补贴额。政府会授标给所标明补偿额最低的农民。[①] 因此,对于澳大利亚雨养农业农民而言,存在通过相对低成本的温室气体减排而赚取额外利润的机会。[②] 需要注意的是,由于澳大利亚雨养农业的温室气体减排成本存在差异性,一些减排成本高于其他行业的雨养农业农民,将无法参与排放减量基金这一减排计划。

2015 年欧盟温室气体配额(EUA)平均价格约为 13.8 澳元/tCO_2e。样本雨养农业农场的温室气体减排成本与其接近,这意味着对于这些家庭农场而言,存在着在排放交易市场上出售排放额度并因此获益的机会。雨养农业能够在减少其温室气体排放的同时以高于其影子价格的金额出售排放额度。而对那些影子价格要高于排放额度价格的行业而言,它们更愿意选择购买排放额度。通过这样的排放额度交易,雨养农业能够获取额外的利润,而那些购买了排放额度的行业能够降低其温室气体排放所需要的成本。为了提高全社会温室气体减排的整体效率,政策制定者应当将雨养农业纳入温室气体减排政策体系中。温室气体减排的总社会成本将会因排放额度交易的进行而减少。理论上,排放额度交易会持续进行,直到各行业的影子价格达到均等。[③] 因此,应当支持将澳大利亚碳市场与欧盟碳市场整合在一起的计划。

应当注意的是,近年来欧盟温室气体配额价格与其他国家碳市场的配额交易价格都有所下降。因此,能够出售排放配额的家庭农场会变少。然而,

---

[①] Clean Energy Regulator. Planning for an auction [R]. Clean Energy Regulator, Canberra, 2016.
[②] Baumber, A., Metternicht, G., Cross, R., Ruoso, L. E., Cowie, A. L., Waters, C. Promoting co-benefits of carbon farming in Oceania: Applying and adapting approaches and metrics from existing market-based schemes [J]. Ecosystem Services, 2019, 39, 100982.
[③] Kumar, S., Managi, S., Jain, R. K. CO_2 mitigation policy for Indian thermal power sector: Potential gains from emission trading [J]. Energy Economics, 2020, 86, 104653.

由于样本家庭农场的影子价格在不同年份间存在波动，预计能够出售排放配额的家庭农场的比例在不同年份间也会有所变化。

4.4.4 进一步讨论

需要注意的是，本章在估计农业温室气体减排潜力时，使用的是实际农场数据。相关分析并没有将消费者饮食结构方面的变化考虑在内。此外，考虑到作物产出占农场总产出的大多数，本章影子价格的测算使用的是作物市场价格。值得注意的是，包括如中国和越南在内的许多新兴经济体消费者的饮食结构中，肉类消费的比重正在不断增加[1]。这样的变化会使得在一定时期内，畜牧产品的供需产生失衡，导致相关产品市场价格的上升。可以预期的是，许多国家的农业部门将进一步扩大畜牧生产规模，以应对上升的市场需求。而对作为全球重要畜牧产品出口国的澳大利亚而言，[2] 这种上升趋势会更加明显，而畜牧产出的比重也将随之上升。更进一步地，考虑到市场价格与影子价格之间的关系，研究期间内由畜牧产品价格衍生出的影子价格可能会高于本章估计的结果。

本章利用方向性距离函数所估计的减排潜力，是研究时期内生产技术条件情况下的技术可行解。也就是说，本章的分析并没有考虑潜在的技术进步。实际上，在澳大利亚的雨养农业部门，近年来有一系列改善畜牧生产的技术得到了广泛推广，具体包括改进饲料可消化性、提高活重增加率以及改善牲畜健康状况等方面的技术。[3] 这些技术的采用能够促进当地畜牧生产技术效率以及生产力的提升。[4] 这也意味着，当雨养农业农民充分利用这些新技术时，来自畜牧生产的温室气体排放可以得到进一步降低。因此，实际温室气

---

[1] Frank, S., Havlík, P., Stehfest, E., et al. Agricultural non-CO_2 emission reduction potential in the context of the 1.5℃ target [J]. Nature Climate Change, 2019, 9 (1): 66 – 72. Tang, K., He, C., Ma, C., Wang, D. Does carbon farming provide a cost-effective option to mitigate GHG emissions? Evidence from China. Australian Journal of Agricultural and Resource Economics, 2019, 63 (3): 575 – 592.

[2] Ghahramani, A., Kingwell, R. S., Maraseni, T. N. Land use change in Australian mixed crop-livestock systems as a transformative climate change adaptation [J]. Agricultural Systems, 2020, 102791.

[3] Bailes, K. L., Piltz, J. W., McNeill, D. M. In vivo digestibility of a range of silages in cattle compared with sheep [J]. Animal Production Science, 2020, 60 (5): 635 – 642.

[4] Mayberry, D., Bartlett, H., Moss, J., Davison, T., Herrero, M. Pathways to carbon-neutrality for the Australian red meat sector [J]. Agricultural Systems, 2019, 175: 13 – 21.

体减排潜力可能要大于本章的估计值。

从本章的分析结果中可以得出几个政策启示。

第一，雨养农业能够在显著减少温室气体排放量的同时节约农业生产所需的投入。经营雨养农业的农民能够通过提高生产效率将其现有生产朝着经济—环境双赢的局面转变，同时实现生产增收以及对气候变化的有效应对。相关政府部门应当考虑改进现有以及设计更多的政策工具，以促进雨养农业减少其生产中的技术无效水平。例如，政府应当修改那些限制雨养农业经营灵活性的不适宜的政策规定。政府也可以考虑为提升雨养农业生产效率以及提升家庭农场经营管理水平，从而制定和实施相应的促进政策和计划。

第二，政策制定者可以使用所测算的影子价格作为不同政策工具中的一个参考值。张和巴朗兹尼（Zhang and Baranzini）指出，优化后的污染物排放税税率应当等于污染物的边际减排成本或影子价格。[①] 因此，若雨养农业被纳入碳税系统中，则所测算的农业温室气体影子价格可以作为碳税税率的一个参考值。政府在筹建排放交易市场时，可以以所测算的影子价格为基准设定排放额度的初始价格。

第三，影子价格的浮动趋势说明，需要为雨养农业设计动态的减排定价政策。如果减排的机会成本无法得到完全补偿，雨养农业农民将不愿进行农业温室气体减排活动。因此，政府可以考虑动态修订碳税税率以及合理地设定温室气体交易市场的排放权初始价格，以反映减排机会成本的波动。

第四，考虑到作物产出比重对于温室气体排放影子价格的潜在影响，雨养农业温室气体减排政策应当体现家庭农场间生产经营特征的异质性。为了降低整体社会减排成本，政府在对雨养农业减排任务进行分解时，需要将不同种类家庭农场多样化的影子价格考虑在内。从事雨养农业生产的家庭农场，其减排任务应当与其排放的影子价格相匹配。

4.5 本章小结

本章使用一个允许在技术可行范围内对投入以及厌恶产出同时进行削减

[①] Zhang, Z., Baranzini, A. What do we know about carbon taxes? An inquiry into their impacts on competitiveness and distribution of income [J]. Energy Policy, 2004, 32 (4): 507–518.

的参数型方向性距离函数,对澳大利亚一个重要的雨养农业系统的家庭农场温室气体减排潜力和成本(影子价格)进行了测算。研究结果显示,经营雨养农业的农民能够通过提高生产效率将其现有生产朝着经济—环境双赢的局面转变,同时实现生产增收以及对气候变化的有效应对。2006~2013年,年均温室气体减排潜力值在14%~33%之间波动,均值为21%。样本农场平均影子价格为17.6澳元/tCO_2e(约合88元人民币/tCO_2e)。总体来看,以上研究结果认为,在雨养农业部门进行温室气体减排活动具有成本有效性上的一定优势。

第5章 澳大利亚雨养农业种—畜复合经营家庭农场对温室气体减排政策的响应

5.1 引 言

众多的科学证据都支持这一观点：减轻气候变化所造成的负面影响就必须减少人类活动所产生的温室气体。① 因此，许多国家都已经制定了有针对性的政策来推动温室气体的减排，并宣布了其温室气体减排国家目标。例如，澳大利亚联邦政府所设定的国家减排目标是：2020年温室气体排放量在2005年基础上减少13%。②

农业是包括二氧化碳、甲烷和氧化亚氮在内的温室气体的重要排放源之一。例如，改变耕作方式所引起的碳流失以及肥料使用和饲养牲畜都会产生温室气体排放。在全球范围内，农业产生了超过13%的人类活动所产生的温室气体。③ 农业是澳大利亚第二大温室气体排放源，2013年其排放量占全国总排放量的16%。④ 因此，农业应当为实现全球及国家温室气体减排目标做出贡献。

---

① IPCC. Climate Change 2014：Mitigation of Climate Change [M]. New York, USA：Cambridge University Press, 2015.

② Australian Government. Setting Australia's post-2020 target for reducing greenhouse gas emissions. Final report of the UNFCCC Taskforce [R]. Commonwealth of Australia, Canberra, 2015.

③ WRI. Climate Analysis Indicators Tool：WRI's Climate Data Explorer [Z]. World Resources Institute, Washington, D. C., 2014.

④ Western Australian Government. How Australia accounts for agricultural greenhouse gas emissions [R]. Western Australian Government, Perth, WA, 2017.

碳汇农业被认为是减少农业温室气体排放量的重要途径。碳汇农业指的是农业土地经营者通过农地利用以及农业生产行为来增加土壤以及植被的固碳量或减少农业生产的温室气体排放量。① 保护性耕作、作物残茬管理、用多年生作物代替一年生作物以及改变牲畜牧放方式都可能增加农地土壤中所存储的碳，减少释放到大气中的碳。② 此外，农民还可以通过采取诸如改变作物—牧草种植结构和优化畜牧管理的碳汇农业措施来减少非二氧化碳温室气体的排放。③

澳大利亚已经制定了相关政策来鼓励农民发展碳汇农业。最早出台的是于2011年12月开始实施的《碳汇农业方案》。该方案是一项自愿性温室气体减排政策。④ 该方案规定，农民能够通过实施碳汇农业来赚取可以在自愿碳汇市场上出售的碳排放信用额。2012年7月~2014年7月，《碳汇农业方案》是以一种碳定价的形式运作的。碳定价为农民在生产活动中所固定的碳或减少的温室气体排放提供了价值参考。从2014年12月开始，《碳汇农业方案》被一项新的政策框架即排放减量基金所替代。⑤ 该基金是以一种保留拍卖的形式运作的：农民受邀投标，投标书中明确了其自愿采取的碳汇农业措施，以及在采取这些碳汇农业措施后固定的每一吨碳或减少排放的每一吨温室气体所希望得到的补贴额。政府会授标给所标明补偿额最低的农民。⑥

尽管澳大利亚政府鼓励发展碳汇农业，但目前学界对于农民面对农业温室气体减排政策将如何反应这一问题却鲜有探讨。本章通过利用一个全农场生物经济模型，分析了澳大利亚农民在面对不同程度的农业温室气体减排政策的情况下，其农地利用模式、农业生产行为以及温室气体排放量将有何变

---

① Smith, P., Martino, D., Cai, Z., et al. Greenhouse gas mitigation in agriculture [J]. Philosophical Transactions of the Royal Society of London B: Biological Sciences, 2008, 363 (1492): 789 - 813.

② Sanderman, J., Farquharson, R., Baldock, J. Soil carbon sequestration potential: a review for Australian agriculture [R]. CSIRO Sustainable Agriculture National Research Flagship, Urrbrae, SA, 2010. Tang, K., Kragt, M. E., Hailu, A., Ma, C. Carbon farming economics: What have we learned? [J]. Journal of Environmental Management, 2016, 172: 49 - 57.

③ Bosch, D. J., Stephenson, K., Groover, G., Hutchins, B. Farm returns to carbon credit creation with intensive rotational grazing [J]. Journal of Soil and Water Conservation, 2008, 63 (2): 91 - 98. Bellarby, J., Tirado, R., Leip, A., Weiss, F., Lesschen, J. P., Smith, P. Livestock greenhouse gas emissions and mitigation potential in Europe [J]. Global Change Biology, 2013, 19 (1): 3 - 18.

④⑤ Parliament of the Commonwealth of Australia. Carbon Farming Initiative Amendment Bill, Explanatory Memorandum [R]. Parliament of the Commonwealth of Australia, Canberra, 2014.

⑥ Clean Energy Regulator. Planning for an auction [R]. Clean Energy Regulator, Canberra, 2016.

化。本章的研究对象是雨养农业区种—畜复合农业，是澳大利亚重要的农业部门（大约涵盖了全国农业从业人口30%）。[1] 此外，通过分析农民对农业温室气体减排政策的响应，本章估算了农业温室气体减排的边际减排成本。希望本书能对设计和制定合理有效的农业温室气体减排政策提供有价值的参考。

5.2 相关研究述评

尽管政策制定者对于碳汇农业表现出浓厚的兴趣，目前学界对于农民面对农业温室气体减排政策时将如何改变其农业生产行为和农地利用决策这一问题却鲜有探讨。已有研究主要集中于对固碳成本的估算。研究显示，保护性耕作[2]、持续性耕作[3]、轮作[4]、作物残茬管理[5]以及退耕还林[6]的固碳能力都比较可观。然而，由于在研究区域、农业系统以及减排措施上存在差异，致使这些措施的成本存在一定差异。总体来看，已有研究显示，在发达国家，增加土壤有机碳含量的农业活动是相对低成本的固碳措施；而在发展中国家，退耕还林的可行性较强。

---

[1] Dumbrell, N. P., Kragt, M. E., Gibson, F. L. What carbon farming activities are farmers likely to adopt? A best-worst scaling survey [J]. Land Use Policy, 2016, 54: 29 – 37.

[2] Grace, P. R., Antle, J., Aggarwal, P. K., Ogle, S., Paustian, K., Basso, B. Soil carbon sequestration and associated economic costs for farming systems of the Indo-Gangetic Plain: A meta-analysis [J]. Agriculture, Ecosystems & Environment, 2012, 146 (1): 137 – 146.

[3] Antle, J. M., Capalbo, S. M., Mooney, S., Elliott, E. T., Paustian, K. H. Economic analysis of agricultural soil carbon sequestration: an integrated assessment approach [J]. Journal of Agricultural and Resource Economics, 2001, 26 (2): 344 – 367.

[4] Tschakert, P. The costs of soil carbon sequestration: an economic analysis for small-scale farming systems in Senegal [J]. Agricultural Systems, 2004, 81 (3): 227 – 253. González-Estrada, E., Rodriguez, L. C., Walen, V. K., et al. Carbon sequestration and farm income in West Africa: Identifying best management practices for smallholder agricultural systems in northern Ghana [J]. Ecological economics, 2008, 67 (3): 492 – 502.

[5] Kragt, M. E., Pannell, D. J., Robertson, M. J., Thamo, T. Assessing costs of soil carbon sequestration by crop-livestock farmers in Western Australia [J]. Agricultural Systems, 2012, 112: 27 – 37.

[6] Stavins, R. N. The cost of carbon sequestration: a revealed-preference approach [J]. The American Economic Review, 1999, 89 (4): 994 – 1009. Hunt, C. Economy and ecology of emerging markets and credits for bio-sequestered carbon on private land in tropical Australia [J]. Ecological Economics, 2008, 66 (2): 309 – 318. Hoang, M. H., Do, T. H., Pham, M. T., van Noordwijk, M., Minang, P. A. Benefit distribution across scales to reduce emissions from deforestation and forest degradation (REDD +) in Vietnam [J]. Land Use Policy, 2013, 31: 48 – 60.

分析澳大利亚种—畜复合农业部门温室气体的边际减排成本是评价相关政策成本有效性的一项必要工作。当温室气体边际减排成本高于政府所提供的减排政策价格（如农业碳税）时，农民不会选择减少温室气体排放。污染控制最小成本理论也认为，边际减排成本低的企业会更多地削减污染物排放量，以便产生排污权余额供出售；而边际减排成本较高的企业则会减少对污染的治理，代之以购买相应的排污权，以减少其污染减排成本。最终，整个社会将达到帕累托最优。① 因此，测算温室气体边际减排成本对评价种—畜复合农业中减排措施的经济可行性而言是十分必要的。

目前，有关澳大利亚雨养农业区种—畜复合农业部门温室气体边际减排成本测算的研究还较少。唐凯等（2016）分析了一些澳大利亚雨养农业区种—畜复合农场的温室气体边际减排成本。研究发现，1998~2005 年平均边际减排成本为 29.3 澳元/tCO_2e。② 值得注意的是，一些研究测算了平均减排成本。扎莫等（Thamo et al.）测算了西澳大利亚小麦带地区的一个种—畜复合农场的平均减排成本。他们认为，平均减排成本将不低于 29.3 澳元/tCO_2e。③

已有研究对于不同种—畜经营结构所带来的温室气体排放量变化这一问题关注得较少。绝大部分探讨碳汇农业对温室气体减排影响效果的文献只关注了种植部门的固碳措施④。然而，畜牧部门是全球人类非二氧化碳温室气体的主要排放源，其甲烷排放量为 31 亿 tCO_2e（占全球人类甲烷排放总量的 44%），氧化亚氮排放量为 20 亿 tCO_2e（占全球人类氧化亚氮排放总量的 53%）。⑤ 在

---

① Fan, Y., Wu, J., Xia, Y., Liu, J. Y. How will a nationwide carbon market affect regional economies and efficiency of CO_2 emission reduction in China? [J]. China Economic Review, 2016, 38: 151 – 166.

② Tang, K., Hailu, A., Kragt, M. E., Ma, C. Marginal abatement costs of greenhouse gas emissions: broadacre farming in the Great Southern Region of Western Australia [J]. Australian Journal of Agricultural and Resource Economics, 2016, 60: 459 – 475.

③ Thamo, T., Kingwell, R. S., Pannell, D. J. Measurement of greenhouse gas emissions from agriculture: economic implications for policy and agricultural producers [J]. Australian Journal of Agricultural and Resource Economics, 2013, 57: 234 – 252.

④ 例如 Antle, J. M., Capalbo, S. M., Mooney, S., Elliott, E. T., Paustian, K. H. Economic analysis of agricultural soil carbon sequestration: an integrated assessment approach [J]. Journal of Agricultural and Resource Economics, 2001, 26 (2): 344 – 367. Skidmore, S., Santos, P., Leimona, B. Targeting REDD +: An Empirical Analysis of Carbon Sequestration in Indonesia [J]. World Development, 2014, 64: 781 – 790.

⑤ IPCC. Summary for policymakers [M]. Cambridge, United Kingdom and New York, USA: Cambridge University Press, 2007.

澳大利亚，畜牧部门温室气体排放量约占全国农业排放量的1/2①。所以在分析过程中，不仅需要考虑作物种植过程中所固定的碳，还有必要将牲畜肠道发酵和粪便所产生的温室气体考虑在内。目前尚无以澳大利亚种—畜复合农业为例分析不同种—畜经营结构所带来的温室气体排放量变化的研究。

总体而言，目前学界对于种—畜复合经营农民面对农业温室气体减排政策时将如何改变其农业生产行为和农地利用决策这一问题鲜有探讨。本书的主要贡献在于，通过利用一个全农场生物经济模型分析了澳大利亚雨养农业区种—畜复合经营农民对农业温室气体减排政策的响应。

5.3 研究对象与方法

5.3.1 研究区域自然环境及雨养农业生产概况

本书构建了一个代表澳大利亚西澳大利亚州大南区（Great Southern Region）农业系统的生物经济学模型。该区域位于西澳大利亚州南部，面积3.9万平方千米，南临南大洋，海岸线为250千米，并向内陆延伸约200千米②。该地区共包括11个行政区。

大南区的气候深受南半球副热带高压脊冬夏运动的影响。在冬季，副热带高压脊北移，驱动由西向东从海洋到陆地的湿润气流吹向该地区。在夏季，副热带高压脊南移，引起由东向西从陆地到海洋的干燥炎热气流覆盖该地区。副热带高压脊冬夏运动使得该地区为典型的地中海式气候，夏季炎热少雨，冬季温和多雨。③ 大部分地区年均降水量为400~500毫米，沿海地带部分地区可达900毫米，内陆部分地区为300毫米，地区平均年降水量约为500毫米。地从沿海地带到内部地区降水量逐步减少，约70%的降水发生在冬春两季（5~10月）。冬季雨量丰沛稳定，夏季降水变化较大。该地区的南部沿海地带夏季降水相较内陆地区发生地更为频繁（见图5-1）。

---

① Kragt, M. E., Pannell, D. J., Robertson, M. J., Thamo, T. Assessing costs of soil carbon sequestration by crop-livestock farmers in Western Australia [J]. Agricultural Systems, 2012, 112: 27-37.
② http://www.gsdc.wa.gov.au/region/geography.
③ Climate Kelpie. Western Australia-weather and climate drivers [R]. Climate Kelpie, 2016.

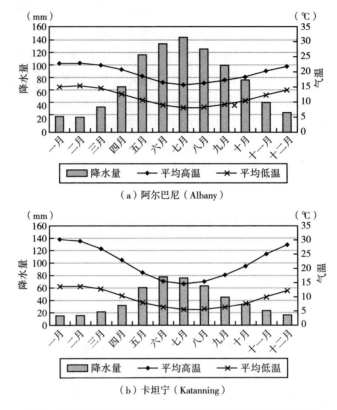

图 5-1　西澳大利亚州大南区部分地区气候状况示意图

注：阿尔巴尼（Albany）位于大南区南部沿海地区，年均降水量为 940 毫米；卡坦宁（Katanning）位于该地区北部，年均降水 480 毫米。本图为作者依据澳大利亚气象局数据绘制而成。

在夏季，由于该地区具有太阳高度大、天空云量少的特征，地表太阳辐射可达 30MJ/m²/天。① 如此高的太阳辐射水平使得当地在夏季处于高温、高蒸发量的状态。此地夏季降水不足，加之高蒸发量，使得作物在夏季生长困难。在冬季，地表太阳辐射约为 10 MJ/m²/天，降雨较多，足以维持作物的生长。

该地区的北半部作物的生长期通常较短，由每年的 4 月持续到当年的 9 月；南半部生长期较长，由 5 月持续到 11 月。每日温差呈现从沿海到内陆逐步升高的态势。该地区气候条件适合一年生作物与牧草的生产。

---

① Kingwell, R. S. Using mathematical programming to model farm management under price and seasonal uncertainty: An analysis of stabilisation policies for wheat and wool [D]. Perth: The University of Western Australia, 1996.

大南区是一个典型的雨养农业区，当地绝大部分农民从事种—畜复合农业经营。① 该地区是澳大利亚的主要农业生产地区之一。近年来，当地谷物平均产量为每公顷 2 吨，同期西澳大利亚州为每公顷 1.4 吨，澳大利亚全国为每公顷 1.3 吨。② 当地许多农场是以种植业为主，超过一半的耕地被用作种植作物。③

目前该地区单个农场的面积从 500 公顷到 5000 公顷不等，均值约为 1900 公顷，存在不同的土壤类型。通常，过半的土地用于作物生产，余下的土地种植牧草供放牧的牲畜食用。④ 主要的农产品包括谷类、活羊与羊毛，绝大多数都被出口到国际市场。⑤

自 20 世纪后期开始，农业生产技术的提升和农业机械化程度的不断提高使得该地区的劳动生产力得到显著提高，单个农场的面积不断扩大。目前，当地的农场通常由农户家庭经营。由于此地区的农业生产已实现了高度的机械化，当地的农场通常由农场主个人经营，往往只雇用一个长期工人。在播种、收割和剪羊毛等农忙季节，一些农户会雇用一定数量的临时劳力。⑥

总体来看，该地区的土壤由于稳定岩石基底的长期风化而肥力较低。⑦ 当地农民通过施用磷酸盐肥料和氮肥来提高作物产量。该地区土壤的其他特点还包括普遍砖红壤化、质地粗以及黏土矿物含量较高。⑧ 该地区农地的土壤主要是浅层、砂质混合土，在其他地区也许被认为并不适宜耕种。本章在生物经济学建模过程中，将当地的土壤概括为四类，并在表 5 - 1 中进行了描述。

---

① Thamo, T., Kingwell, R. S., Pannell, D. J. Measurement of greenhouse gas emissions from agriculture: Economic implications for policy and agricultural producers [J]. Australian Journal of Agricultural and Resource Economics, 2013, 57: 234 - 252.

② Kragt, M. E., Pannell, D. J., Robertson, M. J., Thamo, T. Assessing costs of soil carbon sequestration by crop-livestock farmers in Western Australia [J]. Agricultural Systems, 2012, 112: 27 - 37.

③④ Tang, K., Hailu, A., Kragt, M. E., Ma, C. Marginal abatement costs of greenhouse gas emissions: broadacre farming in the Great Southern Region of Western Australia [J]. Australian Journal of Agricultural and Resource Economics, 2016, 60: 459 - 475.

⑤ Doole, G. J., Bathgate, A. D., Robertson, M. J. Labour scarcity restricts the potential scale of grazed perennial plants in the Western Australian Wheatbelt [J]. Animal Production Science, 2009, 49: 883 - 893.

⑥ Addai, D. The economics of adaption to climate change by broadacre farmers in Western Australia [D]. Ph. D. thesis, School of Agricultural and Resource Economics, University of Western Australia, 2013.

⑦ Moore, G. (ed.). Soil guide: A handbook for understanding and managing agricultural soils [R]. Department of Agriculture, Western Australia. Bulletin No. 4343, 2001.

⑧ Singh, B. Mineralogical and chemical characteristics of soils from south-western Australia [D]. Ph. D. thesis, School of Agriculture, University of Western Australia, 1991.

表 5-1　　　　　　　　　　　模型中包括的土壤种类

土壤种类	特征描述
灰白色沙土	沙土深度超过 80 厘米；顶部 30 厘米为白色、灰色或浅黄色；pH 中性至酸性；可能出现砾石；深部可能出现咖啡石（黏结沙土）
栗色/棕色沙土	沙土深度超过 80 厘米；顶部 30 厘米为黄色/棕色；pH 中性至酸性；可能有腐殖质；可能出现砾石；覆盖在岩石、黏土或其他疏松物质上
混合土	灰色砂质壤土、壤质砂土、砾质砂土和砂土覆盖在有黄色或红色斑点的白色黏土上；酸性
壤土/黏土	顶部土壤呈红色或红棕色；pH 多为中性；表层质地较硬

小麦（triticum aestivum）是谷物生产的大宗，20 世纪 70 年代后期以来，该地区引进了新的作物进行种植。① 羽扇豆（lupinus angustifolius）先与小麦进行轮作。② 一部分收获后的羽扇豆会被储存起来，作为秋季饲料短缺期牲畜的补充饲料。20 世纪 90 年代后期开始，油菜（brassica napus）也逐步被引入当地，在该地区降水较为丰沛的区域有一定的种植规模。近十年来由于相对产量的提高以及市场价格的走高，该地区的农民进行了大规模的油菜种植。豆科作物如鹰嘴豆（cicer arietinum）和紫花豌豆（pisum sativum）通常种植于壤土/黏土区域。

对于大多数作物而言，磷酸盐肥料和氮肥的施用可以有效地提高产量。作物种植过程中广泛使用了化学除草剂以控制杂草生长。谷类作物在完成收购后，往往由农民或承包商运到农场内或其他地点的筒仓内进行储存。部分收获的谷物被用作种子，还有部分的谷物被用作牧草供应有限时牲畜（主要是绵羊）的辅助饲料。③

在该地区，豆科牧草与其他农作物进行轮作。牧草种植的种类因土壤种类与品质的差异而有所不同。在深沙土和混合土区域，普遍种植的牧草种类包括黄花鸡足豆（ornithopus compressus）、地三叶（trifolium subterraneum）、香草、自生一年生牧草以及当地豆类；在壤土/黏土区域，南苜蓿（medicago polymorpha）是主要的品种。④ 近年来，多年生植物如油桉（eucalyptus kochii）

---

①③ John, M. The economics of dryland salinity management in a low-rainfall environment of Western Australia [D]. Ph. D. thesis, School of Agricultural and Resource Economics, University of Western Australia, 2004.

② Marsh, S. P., Pannell, D. J., Lindner, R. K. Does agricultural extension pay? A case study for a new crop, lupins, in Western Australia [J]. Agricultural Economics, 2004, 30 (1): 17-30.

④ Addai, D. The economics of adaption to climate change by broadacre farmers in Western Australia [D]. Ph. D. thesis, School of Agricultural and Resource Economics, University of Western Australia, 2013.

也开始被引入当地。①

牧草生物量的数量主要受到天气条件、土壤种类、放牧压力和肥料等因素的影响。牧草的生产通常始于秋季末期或初冬的初雨时,在春季时牧草生物量达到顶峰,在夏初(10月和11月)时开始枯萎,但仍可作为一段时期内牲畜的饲料。此外,豆科牧草所固定的氮有助于在其后种植的作物的生长。

该地区绝大多数的农户都饲养绵羊(美利奴绵羊和杂交种羊)。羊群可主要用作培育羔羊、培育肥羔、出口活羊、出产羊毛等。该地区绝大部分的绵羊种群保持自然繁育状态。美利奴(Merino)绵羊是当地最主要的牲畜,其体格较大,所产羊绒的纤维直径为20~22微米。② 每只成年羊的羊绒通常重4~6千克。

绵羊通常在5~10月被牧放在种植了牧草的土地上。12月到4月,绵羊的饲料主要是牧草残茬和作物秸秆。2~5月,当地常常出现饲料供给短缺的状况。此时农民会用羽扇豆和其他谷物作为维持羊群营养的补充饲料。农民有时也会种植紫花苜蓿(medicago sativa)作为补充饲料以降低谷物饲养的成本。

农场牧放羊群的结构取决于羊毛与羊肉的相对价格、幼年阉羊的实地交易价以及不同类别羊的饲养成本。③ 秋末和冬季产羔,春季和秋季剪羊毛。所有羊羔的尾巴都被割掉,雄性羊羔都被阉割。幼年阉羊以活畜的形式出口;母羊会留下来用作生产羊毛以及产羔,最终其会被屠宰并以羊肉的形式出售。

5.3.2 农业生产活动建模

本章的分析将主要利用全农场生物经济模型卡坦宁(katanning)进行。④ 基于全农场的视角构建生物经济模型被认为是检验农业系统中不同部分间复

---

①④ Hailu, A., Durkin, J., Sadler, R., Nordblom, T. L. Agent-based modelling study of shadow, saline water table management in the Katanning catchment, Western Australia [R]. Research Report for RIRDC Project No. PRJ – 000578, 2011.

② Kingwell, R., Jeanne, R. M., Hailu, A. A longitudinal analysis of some Australian broadacre farms' greenhouse gas emissions, farming systems and efficiency of production [J]. Agricultural Systems, 2016, 146: 120 – 128.

③ John, M. The economics of dryland salinity management in a low-rainfall environment of Western Australia [D]. Ph. D. thesis, School of Agricultural and Resource Economics, University of Western Australia, 2004.

杂的相互作用以及优化农业生产活动决策的一个有效方法。[①] 卡坦宁模型是一个综合考虑了雨养种—畜复合农业在生物、技术、财务和管理等方面因素，基于农场层次的动态优化农业模型。模型中涵盖了几百种农业活动，包括每一个农地管理单元上的作物—牧草轮作、牲畜饲料的供给与使用、牲畜的繁殖、现金流的记录、农机成本与间接开销等。模型的目标函数是对将农场收入扣除变动成本以及间接费用后所得到的毛利进行最大化。模型关注的是考虑农地管理单元利用历史（或作物轮作）效果的优化后的农地利用序列选择。目前，该模型考虑的农地管理单元利用历史时长为两年。一段时期内的优化过程考虑该时期前的农地管理单元利用历史以及该时期内作物/牧草种植的顺序影响。模型的最优解描述的是基于农场可利用资源可产生最大化毛利润的一系列农场生产活动的一个集合。模型包括了一系列的约束条件，可分为资源约束、逻辑约束、技术约束以及其他方面约束等。

考虑到雨养农业生产的实际，本章拟采用截尾正态随机前沿生产函数的二次形式来估计作物的产量。这将允许在生产过程中存在无效，即产量可以低于既定生产技术条件下的最优值，以及过量农业投入（如发生过量施肥和降水过多等）导致产量递减的情况发生。作物产量由施肥量、降水量、生产随机前沿的噪音项和技术无效来决定，其中技术无效被假设为是一个关于土地和农户特征的函数。该函数也将包含一个时间趋势变量来分析生产技术进步的效果。综上，作物产量可简单表示为：

$$c_i = \alpha_{0_i} + \alpha_{N_i} N_i + \alpha_{N_i N_i} N_i^2 + \alpha_{R_i} R_i + \alpha_{R_i R_i} R_i^2 + \alpha_{N_i R_i} N_i R_i + v_i - u_i \quad (5-1)$$

其中，c_i 是第 i 种作物的产量，N_i 是第 i 种作物的施肥量，R_i 是降水量，v_i 是生产随机前沿的噪音项，u_i 表示技术无效，且技术无效 u_i 被假设为是一个关于土地和农户特征的函数。

本章所利用的卡坦宁模型模拟的是澳大利亚大南区一个土地面积为 1934 公顷、从事种—畜复合经营的家庭农场。模型涵盖了 12 个农地管理单元（见表 5-2），包括一系列的作物和牲畜模块。农作物类型包括小麦、油菜、紫花豌豆以及羽扇豆，这些作物与牧草进行轮作；主要的牲畜是美

[①] Bonesmo, H., Skjelvåg, A. O., Henry Janzen, H., Klakegg, O., Tveito, O. E. Greenhouse gas emission intensities and economic efficiency in crop production: A systems analysis of 95 farms [J]. Agricultural Systems, 2012, 110: 142-151. Robertson, J., Pannell, J., Chalak, M. Whole-farm models: a review of recent approaches. Australian Farm Business Management Journal, 2012, 9 (2): 13-26.

利奴绵羊，用作生产羊毛和羊肉；年均降雨量设置为505毫米，反映了当地典型的降水特征。①

表5-2　　　卡坦宁全农场生物经济模型中包括的农地管理单元

编号	面积（公顷）	土壤类型
1	42	灰白色沙土
2	55	栗色/棕色沙土
3	67	栗色/棕色沙土
4	200	混合土
5	160	混合土
6	225	混合土
7	220	混合土
8	138	混合土
9	280	混合土
10	300	混合土
11	225	混合土
12	22	亚黏土/黏土

模型中作物、活羊以及羊毛的价格见表5-3。小麦与油菜的价格使用当地主要种植的澳大利亚优等小麦（australian premium wheat）以及耐三嗪类除草剂油菜（triazine tolerant canola）的价格来表示。所有的价格为农户出售价，单位为2015年澳元。模型假设农民依据作物与牲畜的预期市场价格、农业生产成本以及当地的环境条件（如土壤类型、生长季节降雨量和作物轮作效应）来做出农业生产决策。

表5-3　　　本章模型中所使用的农产品价格

农产品	价格
小麦（澳元/吨）	280
羽扇豆（澳元/吨）	317
紫花豌豆（澳元/吨）	320
油菜（澳元/吨）	480

---

① Holper, P. N. Climate Change, Science Information Paper: Australian Rainfall: Past, Present and Future [R]. CSIRO, 2011.

续表

农产品	价格
羊毛（澳元/千克）（含脂原毛）	7.43
羔羊（澳元/千克）	3.63
出生 12~24 个月母羊（澳元/千克）	1.92
母羊（澳元/千克）	1.89
阉羊（澳元/千克）	1.33

注：作物商品价格引自谷物研究与发展公司（Grains Research and Development Corporation, 2015）。牲畜商品价格引自澳大利亚肉类与畜牧业协会（Meat and Livestock Australia, 2016）。

基于研究目的，本书对卡坦宁全农场模型进行了如下改进。

第一，基于当地的实际情况，对模型中牧业生产的相关参数进行了更新。这些参数包括：不同月份活羊的体重和羊的胴体重占比；母羊在各月中的怀孕比重、产羔比重、处于泌乳期的比重；母羊怀孕单胎比重；羊羔存活比重；不同性别和年龄的绵羊在各月的存活比重、能量需求以及原毛产量。为了验证参数的有效性，笔者咨询了熟悉该地区农业生产系统的农业经济学家、地方农场管理专家以及一位当地的农业咨询师。[①] 这些专家都认为，通过与他们的研究结果及实际经验相比较，模型中所使用的参数是合适的，模型结果符合当地实际情况。

第二，本书为最新版的模型增加了测算农业温室气体年排放量的模块。所使用的测算方法来自澳大利亚政府公布的《国家排放清单报告（2012）》[*National Inventory Report*（2012）]，[②] 并且对相应参数进行了调整以反映当地的农业系统特征。本书考虑了四类农业排放源，包括作物秸秆、化肥施用、牲畜（包括肠道发酵和牲畜粪便）以及固氮作物。所有的温室气体排放量参考唐凯等（2016）使用全球暖化潜能值转化为 CO_2e。[③]

第三，本书还增加了农业温室气体减排政策模块。本书所考虑的农业温室气体减排政策是一种农业碳税，即农民需要为其在生产中所排放的每一吨

[①] 这些专家包括来自西澳大利亚州农业与食品厅的官员、来自西澳大学农业与资源学院的学者以及来自大南区的一位农业咨询师。

[②] 具体可参见 https：//unfccc.int/files/national_reports/annex_i_ghg_inventories/national_inventories_submissions/application/zip/aut-2012 – nir-12apr.zip.

[③] Tang, K., Hailu, A., Kragt, M. E., Ma, C. Marginal abatement costs of greenhouse gas emissions: broadacre farming in the Great Southern Region of Western Australia [J]. Australian Journal of Agricultural and Resource Economics, 2016, 60: 459 – 475.

温室气体缴纳一定金额的税。此项碳税不包括免税排放额,且税率固定,不随着温室气体排放量的变化而浮动。碳税的税率从 0 上浮到 100 澳元/tCO_2e,每次上浮的幅度为 10 澳元/tCO_2e。

5.4 研究结果

本章先给出的是没有农业碳税时的模拟结果。接下来展示的是不同农业碳税税率情况下的模拟结果。本章的模拟周期是 10 年。

5.4.1 零农业碳税情景

当农民不用为排放的温室气体支付碳税(即税率为 0)时,通过优化农场的种—畜生产经营结构所得到的最大毛利年均为每公顷 613 澳元(见图 5-2)。此时约有 54% 的农地用作种植作物,余下土地用作种植牧草供羊群食用(见图 5-3)。平均而言,约有 46.2%(894 公顷)、24.8%(479 公顷)、23.7%(458 公顷)和 5.3%(104 公顷)的农地被分别用作种植牧草、小麦、油菜和羽扇豆,而在此情景中紫花豌豆则未被包括在最优种—畜生产组合中。主要的作物轮作组合包括持续性牧草、牧草—油菜—小麦组合、小麦—牧草组合以及油菜—牧草组合。需要注意的是,在该模型所包含的 12 个农地管理单元上所选择的种—畜生产活动因为土壤的差异性而各有所不同,这些种—畜生产活动从总体上保证了整个农业生产系统的毛利润最大化。小麦、油菜和羽扇豆的平均产量分别为每公顷 3.16 吨、每公顷 2.3 吨和每公顷 2.32 吨。以上各模拟结果与已有的关于西澳大利亚地区类似农业系统的研究结论基本一致[1]。

农场温室气体年均排放总量为 4168 tCO_2e(每公顷 2.16 tCO_2e)。在 10 年的分析周期内,农场所养殖的牲畜是最大的温室气体排放源,每年平均排放 3543.1 tCO_2e,约占农场总排放量的 85%(见图 5-3)。固氮类作物(主

---

[1] Schirmer, J., Parsons, M., Charalambou, C. Gavran, M. Socio-economic impacts of plantation forestry in the Great Southern region of WA [R]. Forest and Wood Products Research and Development Corporation, Melbourne, VIC, 2005. Holper, P. N. Climate Change, Science Information Paper: Australian Rainfall: Past, Present and Future [R]. CSIRO, 2011.

要是牧草)则是第二大温室气体排放源,约占农场总排放量的13%。化肥施用以及作物秸秆所产生的排放则相对较少。

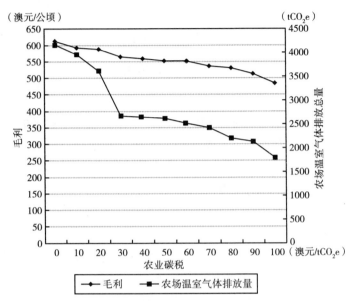

图 5-2　在不同农业碳税情景下的农场毛利与温室气体总排放量

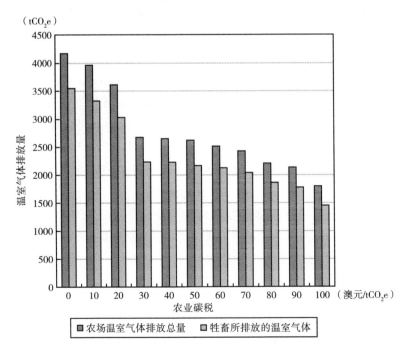

图 5-3　不同农业碳税情景下农场温室气体排放总量与牲畜所排放的温室气体

5.4.2 正农业碳税情景

当农业碳税被设定为 10 澳元/tCO_2e 时,最大毛利年均为每公顷 592 澳元。此时,牧草种植面积、主要作物轮作组合以及作物产量与零农业碳税情景时类似,但油菜的种植面积有所下降(见图 5-4)。小麦依旧是主要的种植作物,与此同时,农民开始将紫花豌豆纳入最优种—畜组合中。农场温室气体年均排放总量为 3958 tCO_2e,与零农业碳税情景相比下降 5%。从图 5-3 中可以看出,所减少的温室气体主要来自牲畜养殖部分。

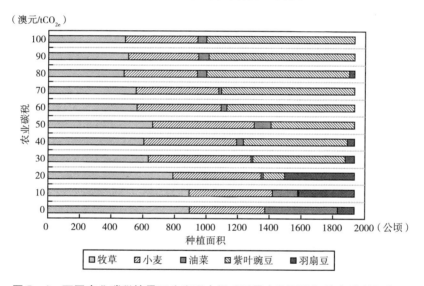

图 5-4 不同农业碳税情景下为实现农场毛利最大化而进行的农地利用分配

当农业碳税被设定为 20 澳元/tCO_2e 时,最大毛利年均为每公顷 588 澳元。此时牧草种植面积平均减少到 788 公顷。农民继续选择减少油菜种植而扩大紫花豌豆的种植,且牧草—紫花豌豆—小麦组合成为重要的作物轮作组合之一(见表 5-4)。农场温室气体年均排放总量显著降低,与零农业碳税情景相比下降了 13.5%。

若农业碳税继续上涨到 30 澳元/tCO_2e,则最大年均毛利减少到每公顷 565 澳元。作物种植面积进一步扩大,小麦和紫花豌豆成为最主要的作物,油菜和羽扇豆的种植面积较小(见图 5-4)。持续性牧草以及牧草—紫花豌豆—小麦组合成为所有农地管理单元上的主要轮作组合。在牧草—紫花豌豆—小麦组合

表5-4　雨养种—畜复合农业在不同农业碳税情景下的优化轮作组合

农业碳税税率（澳元/tCO$_2$e）	轮作组合			
	灰白色沙土	栗色/棕色沙土	混合土	亚黏土/黏土
0	PCW, WPP, WLPC	PPC, PCW, PPW, PPP	PPP, PCW, CPCW, PPW, WPP, PPC, PCP, WPC, WLPC	PPW, PLW
10	PPP, PLW	PPP, PCW	PPP, PCW, PWP, WPCW, WPC, CPPC, LWP, WPL, CLP	PCW, CPCW
20	CPCW, PLW, PPL	PLW, PPL	PPP, PCW, PLW, PWP, PFW, PFWF, PWL, PPW, PLWF, LPLW	PCW, PLW
30	PFW, FPFW	PFW, PPP, PCW	PPP, PCW, PLW, PFW, PFWF, PWL, PWF, PWW, FCW	PLW, PFW, PWF
40	PFW, PWF, WPFW	PFW, PWW	PFW, PFWF, PWF, FPFW, PWW, FCW, PFC, FPF	PFW, FPFW
50	PFWF, PWF, PWW	PCW, PWF, FCW, PCWF	PFW, PFWF, PWF, PWW, PCW, PPW, PFWP	PFW, PWW
60	PFW, PFWF, PWF	PFW, PWF, PWW	PFW, PFWF, PWF, PWW, WFC, PFWFC	FPFW, PWW, FPP
70	PFW, PFWF, PWF	PFWF, CFC	PFW, PFWF, PWF, FPFW, PWW, PPP	PWF, FCW
80	PFWF	PFW, PFWF, PWF	PFW, PFWF, PWF, FCFP, FPFW, PWWF, CFP, CWF, PCF	PFW, PFC
90	PFW, PFWF, PWF	PFW, PFWF	PFW, PFWF, PWF, FCFP, PCF	WFPF, PFCF
100	PFWF, PFC	PFW, PFWF, PWF, FCFP, PCFC	PFW, PFWF, PWF, PCF, PCFC	PFW, PFWF, PWF

注：P=牧草；W=小麦（Triticum aestivum）；C=油菜（Brassica napus）；L=羽扇豆（Lupinus angustifolius）；F=紫花豌豆（Pisum sativum）。

中，小麦的产量可达到每公顷3.5吨，相较其他轮作组合中的产量增加11%。农场温室气体年均排放总量进一步减少，为2671tCO$_2$e，与零农业碳税情景相比下降了35.9%。作物种植在优化后的农业生产系统中所占的比重进一步增加。最优轮作组合中包含油菜的轮作组合进一步减少，而包含紫花豌豆的轮作组合被广泛采用。

若农业碳税在30澳元/tCO_2e基础上继续上涨，则最大年均毛利、牧草种植面积以及农场温室气体年均排放总量的下降幅度开始减小。小麦和紫花豌豆依旧为最主要的作物，所有的农地管理单元均选择牧草—紫花豌豆—小麦组合作为主导性的作物组合。农民倾向于等比例地减少牲畜养殖和作物生产所产生的温室气体（见图5-3）。如果农业碳税被设定为100澳元/tCO_2e，则农场温室气体年均排放总量减少到1795 tCO_2e，与零农业碳税情景相比下降了56.9%。此时最大年均毛利减少到每公顷484.9澳元。过半的农地被用作种植紫花豌豆，余下的农地大致平均地用作种植小麦和牧草。

5.4.3 敏感性分析

为了进一步探索以上分析结果的敏感性，本书进一步模拟在六种不同作物价格情景中被优化的雨养种—畜复合农业系统。优化后的农地利用决策详见图5-5～图5-10。

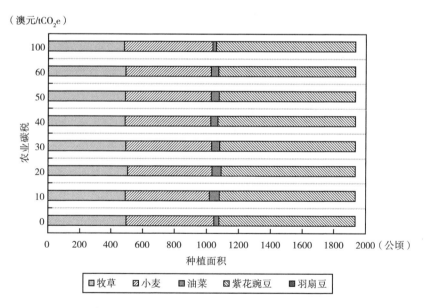

图5-5 小麦价格上涨10%时不同农业碳税情景下最优农地利用分配

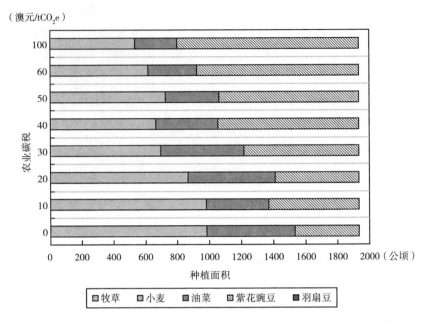

图 5-6 小麦价格下跌 10% 时不同农业碳税情景下最优农地利用分配

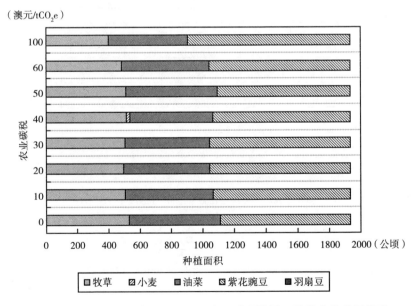

图 5-7 油菜价格上涨 10% 时不同农业碳税情景下最优农地利用分配

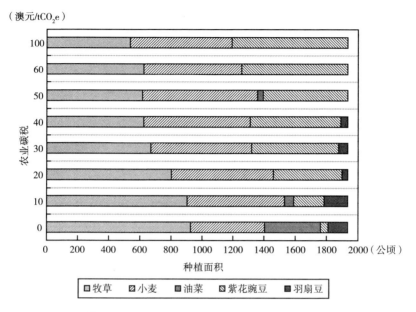

图 5-8　油菜价格下跌 10% 时不同农业碳税情景下最优农地利用分配

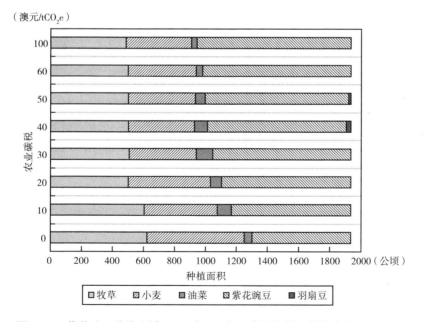

图 5-9　紫花豌豆价格上涨 10% 时不同农业碳税情景下最优农地利用分配

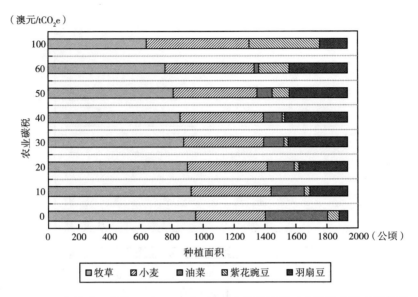

图 5-10　紫花豌豆价格下跌 10% 时不同农业碳税情景下为最优农地利用分配

敏感性分析的结果显示,当农业碳税税率不断提高时,经过优化的雨养种—畜复合农业系统的结构会朝着作物主导型发展,农民倾向于在种植作物时减少包含油菜轮作组合的使用频率而更多地选择包含紫花豌豆轮作组合,以减少温室气体的排放。

图 5-11 中的曲线反映了在七种作物价格情景中不同农业碳税税率下的最优农场温室气体总排放量。可以看到,基准价格情景中的农场温室气体排放水平处于高作物价格情景与低作物价格情景的排放水平之间。当作物价格上涨时,农场温室气体排放水平会较基准价格情景所减少;而当作物价格下跌时,农场温室气体排放水平会高于基准价格情景时的排放量。在雨养种—畜复合农业中,农民面对高作物价格水平会选择扩张作物生产,当作物价格水平偏低时会扩张畜牧生产。考虑到单位面积畜牧生产所排放的温室气体要远高于作物生产所产生的排放,① 扩张畜牧生产会导致农场温室气体排放水平的上升,而扩张作物生产则会降低农场温室气体的排放水平。因此,农场温室气体排放水平会在高作物价格情景中有所下降,而在低作物价格情景中有所上升。

---

① Thamo, T., Kingwell, R. S., Pannell, D. J. Measurement of greenhouse gas emissions from agriculture: Economic implications for policy and agricultural producers [J]. Australian Journal of Agricultural and Resource Economics, 2013, 57: 234-252.

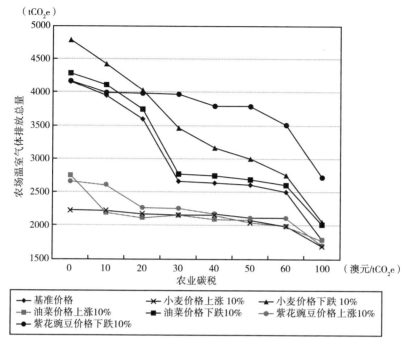

图 5-11 雨养种—畜复合农业在不同作物价格情景以及农业碳税情况下的农场温室气体总排放量

从图 5-11 中还可以看出，在紫花豌豆价格偏低的情况下，即便农业碳税税率较高，最优农场温室气体总排放量仍会处于较高水平。在雨养种—畜复合农业系统中，单一粮食或油料作物的持续性耕作会造成土壤中有机质含量的减少，因而会损害土壤肥力。① 因此，即便农业碳税税率较高，粮食或油料作物仍需要与豆科作物（如豆科牧草和紫花豌豆）进行轮作以确保土壤肥力和作物产量。在低紫花豌豆价格情景中，畜牧生产相较于作物生产特别是紫花豌豆生产所获得的利润更高，农民会选择扩大畜牧生产规模，种植更多的牧草作为牲畜饲料，同时还可以保持土壤的肥力。考虑到单位面积畜牧生产所排放的温室气体要高于作物生产所产生的排放，此时即便农业碳税税率较高，最优农场温室气体总排放量仍会处于较高水平。

---

① Reeves, D. W. The role of soil organic matter in maintaining soil quality in continuous cropping systems [J]. Soil and Tillage Research, 1997, 43 (1): 131-167.

5.5 进一步讨论

本章通过利用一个全农场生物经济模型,分析了澳大利亚雨养种—畜复合经营农民在面对不同程度的农业温室气体减排政策的情况下,其农地利用模式、农业生产行为以及温室气体排放量将有何变化。研究结果显示,当农业碳税税率不断提高时,经过优化的雨养种—畜复合农业系统的结构会朝着作物主导型发展,从而减少温室气体的排放。在雨养种—畜复合农业系统中,牲畜养殖相较作物生产通常产生更多的温室气体。[①] 在不同农业碳税税率的情景中,牲畜养殖所产生的温室气体占到了农场排放总量的83%~84%(见图5-3)。牲畜生产所产生的温室气体约为每公顷3.6tCO_2e,大致是每公顷作物生产所产生温室气体排放(0.37tCO_2e)的10倍。考虑到牲畜生产由于农业碳税的实施所产生的外部成本较高,农民不得不限制包括养殖绵羊和种植牧草在内的畜牧产业,提高雨养种—畜复合农业系统中作物生产的比重。

以上分析结果还显示,当农业碳税税率不断提高时,农民倾向于在种植作物时减少包含油菜的轮作组合的使用频率,而更多地选择包含紫花豌豆的轮作组合。在半干旱条件下,作物所产生的温室气体与其在生产过程中所需施用的氮肥数量以及作物器官的氮浓度密切相关。[②] 已有研究表明,包含油菜的轮作组合与包含紫花豌豆的轮作组合相比,前者在生长过程中需要更多的氮肥,因而会排放更多的温室气体。[③] 紫花豌豆根部共生的豌豆根瘤菌具有固氮作用,其固定的氮素能满足紫花豌豆生长期所需量的2/3左右,因而种植紫花豌豆不需要大量的氮肥。此外,紫花豌豆茬的土壤肥力比麦类茬高。由于紫花豌豆的生物固氮作用,其根和残茬的碳氮比低,分解速度更快,提

[①] Thamo, T., Kingwell, R.S., Pannell, D.J. Measurement of greenhouse gas emissions from agriculture: Economic implications for policy and agricultural producers [J]. Australian Journal of Agricultural and Resource Economics, 2013, 57: 234-252.

[②] Gan, Y.T., Campbell, C.A., Jansen, H.H., et al. Carbon input to soil by oilseed and pulse crops in semiarid environment [J]. Agriculture, Ecosystem & Environment, 2009, 132: 290-297.

[③] Gan, Y., Liang, C., Hamel, C., Cutforth, H., Wang, H. Strategies for reducing the carbon footprint of field crops for semiarid areas. A review [J]. Agronomy for Sustainable Development, 2011, 31 (4): 643-656.

高了有机质的含量水平，增加了土壤微生物量以及矿化的氮，促进了后茬作物的生长。此外，紫花豌豆与其他作物轮作，可以使后作的产量提高。[①] 例如，豆茬麦与重茬麦相比，可增产一到两成。[②] 进一步地，如果增加轮作中紫花豌豆的种植频率，降低土壤中氮残留量过高对于固氮的负面影响，以及提高来自紫花豌豆茬中氮的矿化与后作氮需求高峰期之间的同步性，后作的产量可以得到进一步提高。[③] 综合上述因素，农民在得到农业温室气体减排政策措施激励的情况下，会倾向于将包含油菜的轮作组合替换为包含紫花豌豆的轮作组合，以减少温室气体的排放，提高经济效益。

进一步地，分析结果显示，随着农业碳税税率的不断提高，农场的利润与温室气体排放总量均会随之降低。但是，在农业税税率不太高的情况下，该农业系统仍可显著减少其温室气体排放总量，且所需成本较小。相较于零碳税情景，农民在农业碳税税率为20澳元/tCO_2e的情况下可以通过优化其种—畜生产组合来减少13.5%的农场温室气体排放，而每公顷的毛利仅减少4%。进一步地，若农业碳税税率为30澳元/tCO_2e，与零碳税情景相比农场温室气体排放可减少36%，而每公顷毛利润减少的幅度低于8%（见图5-2）。

如上所述，当农业碳税的税率高于农业温室气体的边际减排成本时，农民会选择减少其农业温室气体的排放。因此，本书的结果意味着减少13%农业温室气体排放的边际减排成本低于20澳元/tCO_2e，而减少36%排放的边际减排成本低于30澳元/tCO_2e。由于澳大利亚的温室气体减排目标是到2020年的排放量比2005年水平下降13%，可以推断西澳大利亚经营种—畜复合农业的农民可以以低于20澳元/tCO_2e（约合100元人民币/tCO_2e）的边际减排成本减少其13%的温室气体排放，以与国家的减排目标相符。

现将以上估计的减排成本与近期发表的关于澳大利亚农业减排的研究所估计的成本进行比较。所测算的结果均利用澳大利亚储备银行所提供的基于CPI的通胀换算器折算成2015年澳元。[④] 由于涉及澳大利亚农业温室气体边

---

① Kirkegaard, J., Christen, O., Krupinsky, J., Layzell, D. Break crop benefits in temperate wheat production [J]. Field Crops Research, 2008, 107 (3): 185–195.

② Holper, P. N. Climate Change, Science Information Paper: Australian Rainfall: Past, Present and Future [R]. CSIRO, 2011.

③ Herridge, D. F., Peoples, M. B., Boddey, R. M. Global inputs of biological nitrogen fixation in agricultural systems [J]. Plant and Soil, 2008, 311 (1–2): 1–18.

④ http://www.rba.gov.au/calculator/annualDecimal.html.

际减排成本的研究较少,我们也纳入了一些测算了平均减排成本的研究。①以上研究中,只有格雷斯等(Grace et al.)和唐凯等(2016)测算了边际减排成本,其他的研究则都是测算了平均减排成本。② 本书所估计的结果是减少13%温室气体排放的边际减排成本低于 20 澳元/tCO_2e(约合 100 元人民币/tCO_2e),远低于已有研究的结论。

已有研究的结果因模拟周期、研究区域、气候类型、碳汇农业措施以及所采用的模型的不同而各异。唐凯等(2016)所研究的区域与本书有一定的重合,所考虑的碳汇农业措施与本书一致。之所以其研究结论与本书有所差异,主要是因为二者所采用的模型不同。唐凯等(2016)采用的是利用农场实际生产数据、基于投入的距离函数分析法,该法并没有考虑农场土地分配与经营管理方式的优化。与此不同的是,本章所利用的模型是一种优化模型,其所测算的结果是基于农民为取得最大化毛利而进行的一系列优化了的农场生产活动。因此,不同的研究方法使得边际减排成本的测算值存在差异。

本书与唐凯等(2016)之间的差异意味着存在进一步显著降低澳大利亚雨养种—畜复合农业温室气体减排成本的可能。如果雨养种—畜复合经营农民采用了能够获得最大化毛利的一系列优化了的农场生产决策,即便已经减少了一定量的温室气体排放(如本书中所涉及的13%),其接下来的减排成本依旧要低于那些既没有采取优化了的农场生产决策又没有减少温室气体排

---

① Flugge, F., Abadi, A. Farming carbon: an economic analysis of agroforestry for carbon sequestration and dryland salinity reduction in Western Australia [J]. Agroforestry Systems, 2006, 68: 181 – 192. Hunt, C. Economy and ecology of emerging markets and credits for bio-sequestered carbon on private land in tropical Australia [J]. Ecological Economics, 2008, 66 (2): 309 – 318. Grace, P. R., Antle, J., Ogle, S., et al. Soil carbon sequestration rates and associated economic costs for farming systems of south-eastern Australia [J]. Australian Journal of Soil Research, 2010, 48: 1 – 10. Kragt, M. E., Pannell, D. J., Robertson, M. J., Thamo, T. Assessing costs of soil carbon sequestration by crop-livestock farmers in Western Australia [J]. Agricultural Systems, 2012, 112: 27 – 37. Thamo, T., Kingwell, R. S., Pannell, D. J. Measurement of greenhouse gas emissions from agriculture: Economic implications for policy and agricultural producers [J]. Australian Journal of Agricultural and Resource Economics, 2013, 57: 234 – 252. Tang, K., Hailu, A., Kragt, M. E., Ma, C. Marginal abatement costs of greenhouse gas emissions: broadacre farming in the Great Southern Region of Western Australia [J]. Australian Journal of Agricultural and Resource Economics, 2016, 60: 459 – 475.

② Grace, P. R., Antle, J., Ogle, S., et al. Soil carbon sequestration rates and associated economic costs for farming systems of south-eastern Australia [J]. Australian Journal of Soil Research, 2010, 48: 1 – 10. Tang, K., Hailu, A., Kragt, M. E., Ma, C. Marginal abatement costs of greenhouse gas emissions: broadacre farming in the Great Southern Region of Western Australia [J]. Australian Journal of Agricultural and Resource Economics, 2016, 60: 459 – 475.

放的农民。因此，政策决策者应当考虑鼓励雨养种—畜复合经营农民优化其农业生产决策，以增加农业生产决策，同时降低温室气体减排成本。

现在将本书的估计成本与澳大利亚一些重要的经济部门的平均减排成本进行比较。平均减排成本数据选自麦肯锡公司（McKinsey & Company，2008）。[①]图5-12显示，本书所估计的成本远低于林业、交通、能源和建筑业的减排成本。这说明，通过改变农地利用模式以及农业生产行为来减少农业温室气体排放是一个相对低成本的减排途径。

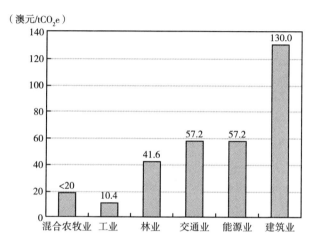

图5-12 澳大利亚不同产业部门温室气体减排成本比较

在解读本书所得出的分析结论时需注意以下问题。第一，读者需要注意，农业温室气体的排放量估计是建立在有关气候条件和土壤类型的一系列假设基础之上。本书的结果是基于处于半干旱地中海式气候条件、以雨养种—畜复合农业为代表的澳大利亚大南区。在其他农业系统和不同气候环境中，农业温室气体减排政策的效果可能存在差异，需要未来进一步研究。第二，本书没有考虑由农业减排活动所带来的潜在的协同效益，如生态多样性的保护以及土壤品质的改良。[②] 然而，农民可能从这些协同效应中受益，从而在实际上降低了温室气体边际减排成本。此外，协同效益产生了公共效用。政策

---

① McKinsey & Company. Australian Cost Curve for GHG Reduction [R]. McKinsey & Company, Australia, 2008. 该报告使用了一个四阶段流程来确定正常经营条件下基准线，而后分析潜在减排的机会与成本。

② Tang, K., Kragt, M. E., Hailu, A., Ma, C. Carbon farming economics: What have we learned? [J]. Journal of Environmental Management, 2016, 172: 49-57.

制定者应当为这些带来了额外益处的农业减排活动提供更高的碳价。第三，本书所探讨的农业温室气体减排政策是一种简单的农业碳排放税，即农民需要为其在生产中所排放的每一吨温室气体缴纳一定金额的税，无免税排放额，且税率固定。对于在不同类型农业温室气体减排政策背景下（如累进制排放税）农民的生产经营行为，还需进一步探讨。第四，采取碳汇农业措施可能会产生一系列与农业生产非直接相关的额外成本，如交易成本和学习成本等。[①] 本书所使用的模型没有考虑这些额外成本。然而在实际操作中，这些额外成本可能给经营种—畜复合农业的农民采取碳汇农业措施带来一定的阻碍。第五，除了气候变动所带来的风险，还需要进一步考虑其他风险对于模拟结果所产生的影响。未来的研究需要进一步地分析和模拟农民风险偏好，以及农产品市场价格变动情况对于农业温室气体减排政策响应所带来的影响。

5.6 本章小结

本章利用一个全农场生物经济学模型，分析了西澳大利亚州大南区的雨养复合农业系统对于程度不同的农业温室气体减排政策响应。其中，农业温室气体减排政策以农业碳税的形式来表示。具体的响应内容包括了在农地利用模式、农业生产行为以及温室气体排放量方面的变化。这是针对澳大利亚雨养种—畜复合经营农民对于农业温室气体减排政策响应所进行的一项新的研究。

研究结果显示，由于牲畜是主要的农业温室气体排放源，优化后的复合农业系统中种植业的比例将上升，以减少农业生产所排放的温室气体。受农业碳税的影响，农民倾向于种植更多的豆科作物而减少油料作物的种植。此外，随着农业碳税税率的增加，农业利润以及农业温室气体排放量均出现下降。然而，相对较低的碳税税率可以实现农业温室气体的有效减排，且减排成本相对较小。分析结果表明，西澳大利亚州农民实现减少13%温室气体排放量国家减排目标的边际减排成本不高于 20 澳元/tCO_2e（约合 100 元人民

---

① Bakam, I., Balana, B. B., Matthews, R. Cost-effectiveness analysis of policy instruments for greenhouse gas emission mitigation in the agricultural sector [J]. Journal of Environmental Management, 2012, 112: 33–44. Tang, K., Kragt, M. E., Hailu, A., Ma, C. Carbon farming economics: What have we learned? [J]. Journal of Environmental Management, 2016, 172: 49–57.

币/tCO_2e),而减少 36% 排放的边际减排成本低于 30 澳元/tCO_2e(约合 150 元人民币/tCO_2e)。

总之,本章的研究结果说明,通过改变农地利用模式以及农业生产行为来减少农业温室气体排放是一个相对低成本的减排途径。澳大利亚雨养种—畜复合经营农民可以通过改变农地利用模式以及农业生产行为,如优化种—畜生产经营结构和作物轮作,以相对较低的成本产生可观的农业温室气体减排。政策制定者需要优先考虑为农业部门推广碳汇农业提供政策支持,以完善减排政策体系。在现有的减排政策如《碳汇农业方案》和排放减量基金的框架内,政策制定者应当考虑为澳大利亚雨养种—畜复合农业提供更有针对性且更具吸引力的减排政策,以提高农民对这些减排政策的参与。

第6章 黄土高原地区雨养农业温室气体减排的成本有效性分析

6.1 引　　言

目前，中国控制温室气体排放面临着来自国际和国内的巨大压力和困难。为了增强适应气候变化的能力，有效控制温室气体过量排放，中国政府对减排目标做出明确承诺：2030年左右中国二氧化碳排放达到峰值，且将努力早日达峰[1]。作为温室气体重要排放源，农业排放了大量的二氧化碳、甲烷和氧化亚氮[2]。例如，作物种植过程产生了碳流失，施用的化肥、养殖的牲畜以及田间焚烧的农业废弃物会产生氧化亚氮和甲烷[3]。在中国，农业排放了约占全国总排放量15%的温室气体，且排放了全国90%的氧化亚氮以及全国60%的甲烷[4]。2010年农业活动的甲烷和氧化亚氮排放量分别为4.71亿tCO_2e和3.58亿tCO_2e[5]。由此可见，农业存在着温室气体减排的巨大潜力，农业

---

[1] 生态环境部：《中华人民共和国气候变化第三次国家信息通报》，2018。

[2] Tang, K., Kragt, M. E., Hailu, A., Ma, C. Carbon farming economics: What have we learned? [J]. Journal of Environmental Management, 2016, 172: 49–57.

[3] 唐凯：《基于生物经济学的澳大利亚农业温室气体减排潜能分析》，人民出版社2018年版。

[4] Tang, K., Hailu, A., Kragt, M. E., Ma, C. Marginal abatement costs of greenhouse gas emissions: broadacre farming in the Great Southern Region of Western Australia [J]. Australian Journal of Agricultural and Resource Economics, 2016, 60: 459–475. Yang, L., Tang, K., Wang, Z., An, H., Fang, W. Regional eco-efficiency and pollutants' marginal abatement costs in China: A parametric approach [J]. Journal of Cleaner Production, 2017, 167: 619–629. Wu, X., Zhang, J., You, L. Marginal abatement cost of agricultural carbon emissions in China: 1993–2015 [J]. China Agricultural Economic Review, 2018, 10 (4): 558–571.

[5] 生态环境部：《中华人民共和国气候变化第三次国家信息通报》，2018。

减排能为中国实现温室气体减排目标作出的贡献不可低估。

碳汇农业被认为是减少农业温室气体排放量的重要途径。碳汇农业是指通过农地利用以及农业生产行为来增加土壤以及植被的固碳量,或减少农业生产的温室气体排放量。[①] 保护性耕作、作物残茬管理、用多年生作物代替一年生作物以及改变牲畜放方式都可能增加农地土壤中所存储的碳,减少释放到大气中的碳。[②] 此外,农民还可以通过采取碳汇农业措施,减少非二氧化碳温室气体的排放,例如,改变作物—牧草种植结构和优化畜牧管理等。[③] 然而,为了激发农民的碳汇农业行为,适当的政策机制通常是不可缺少的。

自20世纪90年代起,中国政府开始采取一系列的措施来应对气候变化。在农业领域,最广为人知的政府项目是开始于1999年的退耕还林工程。自1999年起,按照"退耕还林(草)、封山绿化、以粮代赈、个体承包"的政策措施,四川、陕西、甘肃3省率先开展退耕还林还草试点,2002年在全国范围内全面启动退耕还林还草工程。20年的持续建设,中央财政累计投入5000多亿元,在25个省(区、市)和新疆生产建设兵团的287个地市2435个县(区)实施退耕还林还草5.15亿亩,占同期全国重点工程造林总面积的2/5,成林面积占全球同期增绿面积的4%以上,4100万农户1.58亿农民直接受益。退耕还林工程已成为世界上资金投入最多、建设规模最大、政策

[①] Smith, P., Martino, D., Cai, Z., et al. Greenhouse gas mitigation in agriculture [J]. Philosophical Transactions of the Royal Society of London B: Biological Sciences, 2008, 363 (1492): 789 - 813. Thamo, T., Kingwell, R. S., Pannell, D. J. Measurement of greenhouse gas emissions from agriculture: economic implications for policy and agricultural producers [J]. Australian Journal of Agricultural and Resource Economics, 2013, 57: 234 - 252. Khataza, R. R., Hailu, A., Kragt, M. E., Doole, G. J. Estimating shadow price for symbiotic nitrogen and technical efficiency for legume-based conservation agriculture in Malawi [J]. Australian Journal of Agricultural and Resource Economics, 2017, 61 (3): 462 - 480.

[②] Sanderman, J., Farquharson, R., Baldock, J. Soil carbon sequestration potential: a review for Australian agriculture [R]. CSIRO Sustainable Agriculture National Research Flagship, Urrbrae, SA, 2010. 张四海、曹志平、张国:《保护性耕作对农田土壤有机碳库的影响》,载《生态环境学报》2012年第2期。Tang, K., Kragt, M. E., Hailu, A., Ma, C. Carbon farming economics: What have we learned? [J]. Journal of Environmental Management, 2016, 172: 49 - 57.

[③] Bosch, D. J., Stephenson, K., Groover, G., Hutchins, B. Farm returns to carbon credit creation with intensive rotational grazing [J]. Journal of Soil and Water Conservation, 2008, 63 (2): 91 - 98. Bellarby, J., Tirado, R., Leip, A., Weiss, F., Lesschen, J. P., Smith, P. Livestock greenhouse gas emissions and mitigation potential in Europe [J]. Global Change Biology, 2013, 19 (1): 3 - 18. Fiala, N. Meeting the demand: an estimation of potential future greenhouse gas emissions from meat production [J]. Ecological Economics, 2008, 67 (3): 412 - 419. Hawkins, J., Ma, C., Schilizzi, S., Zhang, F. China's changing diet and its impacts on greenhouse gas emissions: an index decomposition analysis [J]. Australian Journal of Agricultural and Resource Economics, 2018, 62 (1): 45 - 64.

性最强、群众参与程度最高的重大生态工程。① 此外，中国政府在一些中西部地区一些省（区、市）还开展了生态家园富民计划，通过整合各类可再生能源技术和生态农业技术，因地制宜推广以沼气、生物质能、太阳能等为重点的各类能源生态模式和工程技术。② 然而，这些早期政策措施多是旨在保护和改善生态环境，通常更多的是考虑维护粮食安全，农业温室气体减排并不是这些政策措施的重点目标。

在"十二五"时期（2011～2015年），中国政府开始落实专门旨在减少农业温室气体排放的政策，以推动国家温室气体减排计划任务的完成。这些政策以节肥技术推广为工作重点，希望通过减量化、再利用、资源化等方式，降低能源消耗，减少污染排放，提升农业可持续发展能力。中央财政设专项支持规模养殖场进行标准化改造，建设贮粪池、排粪污管网等粪污处理配套设施，降低畜牧业温室气体排放。自2015年开始，中国通过大力发展节水农业、实施化肥零增长行动、实施农药零增长行动、推进养殖污染防治、深入开展秸秆资源化利用等系列行动，控制面源污染和温室气体排放，增强农业应对气候变化的能力。③

除了以上政策措施，中国政府也为农业部门设定了具体的气候变化应对目标。例如，农业部制定《到2020年化肥使用量零增长行动方案》，明确提出以下目标任务：2015～2019年，逐步将化肥使用量年增长率控制在1%以内，力争到2020年，主要农作物化肥使用量实现零增长。优化施肥结构，到2020年，氮、磷、钾和中微量元素等养分结构趋于合理，有机肥资源得到合理利用；测土配方施肥技术覆盖率达到90%以上；畜禽粪便养分还田率达到60%、提高10个百分点；农作物秸秆养分还田率达到60%、提高25个百分点。改进施肥方式，到2020年，盲目施肥和过量施肥现象基本得到遏制，传统施肥方式得到改变；机械施肥占主要农作物种植面积的40%以上、提高10个百分点；水肥一体化技术推广面积1.5亿亩、增加8000万亩。提高肥料利用率，从2015年起，主要农作物肥料利用率平均每年提升1个百分点以上，力争到2020年，主要农作物肥料利用率达到40%以上。④《中华人民共和国

---

① https://www.yicai.com/news/100682133.html.
② http://www.gov.cn/jrzg/2005-10/18/content_79301.htm.
③ 生态环境部：《中华人民共和国气候变化第三次国家信息通报》，2018。
④ 农业部：《农业部关于印发〈到2020年化肥使用量零增长行动方案〉和〈到2020年农药使用量零增长行动方案〉的通知》，2015。

国民经济与社会发展第十三个五年规划纲要》《国家适应气候变化战略（2013－2020年）》《"十三五"控制温室气体排放工作方案部门分工》明确要求，"十三五"期间（2016~2020年）累计减排11亿 tCO_2e 以上，减少农田氧化亚氮排放；到2020年实现单位国内生产总值碳排放下降18%，农田氧化亚氮排放达到峰值，作物水分利用效率提高到 $1.1kg/m^3$ 以上，农村劳动力实用适应技术培训普及率达到70%。以上这些政策目标基本上是通过直接财政支持以及命令控制型规制来实现的。

近年来，中国政府也开始试行市场型温室气体减排政策。人们普遍认为，市场型减排政策与传统的命令控制型政策相比，更加灵活且在成本有效性上更具优势。[①] 然而，无论是2012年开始的区域碳市场试点，还是最近筹划的全国碳交易市场，都没有将农业部门涵盖在内。

尽管中国政府对鼓励碳汇农业以及促进农业温室气体减排表现出强烈的兴趣，目前在中国碳汇农业减排成本有效性分析以及农民对市场型农业减排政策的响应方面还缺乏实证研究。

本章利用全农场生物经济模型，对中国不同强度市场型农业减排政策所引起的农地使用、农户经营行为以及农户温室气体排放的变化进行分析。本章研究关注的是中国最大的雨养农业区以及第二大农业区——黄土高原。[②] 黄土高原是典型的半干旱种—畜复合经营雨养农业区，当地生态环境十分脆弱。本章也通过分析种—畜复合经营农民对市场型农业减排政策的响应，以及测算相应的温室气体减排量，来估算农业温室气体的边际减排成本。希望通过本章的研究，为设计和实施具有成本有效性优势的农业减排政策提供新的参考。

[①] Yang, L., Tang, K., Wang, Z., An, H., Fang, W. Regional eco-efficiency and pollutants' marginal abatement costs in China: A parametric approach [J]. Journal of Cleaner Production, 2017, 167: 619-629. 孙建飞、郑聚锋、程琨等：《面向自愿减排碳交易的生物质炭基肥固碳减排计量方法研究》，载《中国农业科学》2018年第23期。Wu, J., Ma, C. 2019. The convergence of China's marginal abatement cost of CO_2: An emission-weighted continuous state space approach [J]. Environmental and resource economics, 2019, 72 (4): 1099-1119.

[②] Liu, J., Liu, M., Zhuang, D., Zhang, Z., Deng, X. Study on spatial pattern of land-use change in China during 1995-2000 [J]. Science in China Series D: Earth Sciences, 2003, 46 (4): 373-384. 田磊：《变化环境下黄土高原水文气候要素数值模拟及未来预测》，西北农林科技大学2019年博士学位论文。

6.2 相关研究述评

在政策制定者们对碳汇农业表现出强烈兴趣的同时，分析农业温室气体减排政策对农地使用和农户经营行为的影响成为新兴的研究话题。有关碳汇农业的研究尤其关注固碳农业活动的有效性以及成本。已有研究发现，包括保护性耕作[1]、轮作[2]、不间断耕作[3]、作物残茬管理[4]以及退耕还林[5]在内的碳汇农业活动能够实现可观的固碳效果。然而，这些活动的成本会随着区域、农业系统以及减缓活动的不同而有所差别。[6] 例如，保护性耕作对于高度工

---

[1] Pendell, D. L., Williams, J. R., Boyles, S. B., Rice, C. W., Nelson, R. G. Soil carbon sequestration strategies with alternative tillage and nitrogen sources under risk [J]. Applied Economic Perspectives and Policy, 2007, 29 (2): 247-268. Khataza, R. R., Hailu, A., Kragt, M. E., Doole, G. J. Estimating shadow price for symbiotic nitrogen and technical efficiency for legume-based conservation agriculture in Malawi. Australian Journal of Agricultural and Resource Economics [J], 2017, 61 (3): 462-480.

[2] González-Estrada, E., Rodriguez, L. C., Walen, V. K., et al. Carbon sequestration and farm income in West Africa: Identifying best management practices for smallholder agricultural systems in northern Ghana [J]. Ecological Economics, 2008, 67 (3): 492-502. Havlík, P., Valin, H., Mosnier, A., et al. Crop productivity and the global livestock sector: Implications for land use change and greenhouse gas emissions [J]. American Journal of Agricultural Economics, 2012, 95 (2): 442-448.

[3] Antle, J. M., Capalbo, S. M., Mooney, S., Elliott, E. T., Paustian, K. H. Economic analysis of agricultural soil carbon sequestration: an integrated assessment approach [J]. Journal of Agricultural and Resource Economics, 2001, 26 (2): 344-367.

[4] Thamo, T., Kingwell, R. S., Pannell, D. J. Measurement of greenhouse gas emissions from agriculture: economic implications for policy and agricultural producers [J]. Australian Journal of Agricultural and Resource Economics, 2013, 57: 234-252. Antle, J. M., Zhang, H., Mu, J. E., Abatzoglou, J., Stöckle, C. Methods to assess between-system adaptations to climate change: dryland wheat systems in the Pacific Northwest United States [J]. Agriculture Ecosystems & Environment, 2018, 253: 195-207.

[5] Stavins, R. N. The cost of carbon sequestration: a revealed-preference approach [J]. The American Economic Review, 1999, 89 (4): 994-1009. Hunt, C. Economy and ecology of emerging markets and credits for bio-sequestered carbon on private land in tropical Australia [J]. Ecological Economics, 2008, 66 (2): 309-318. Hoang, M. H., Do, T. H., Pham, M. T., van Noordwijk, M., Minang, P. A. Benefit distribution across scales to reduce emissions from deforestation and forest degradation (REDD+) in Vietnam [J]. Land Use Policy, 2013, 31: 48-60.

[6] Hunt, C. Economy and ecology of emerging markets and credits for bio-sequestered carbon on private land in tropical Australia [J]. Ecological Economics, 2008, 66 (2): 309-318. 张凡、李长生:《气候变化影响的黄土高原农业土壤有机碳与碳排放》，《第四纪研究》2010年第3期。Tang, K., Kragt, M. E., Hailu, A., Ma, C. Carbon farming economics: What have we learned? [J]. Journal of Environmental Management, 2016, 172: 49-57.

业化的地区而言也许是一个具有成本有效性优势的选择，但尚处于工业化进程中的地区似乎更倾向于选择退耕还林。①

近年来，开始有学者对中国碳汇农业的有关问题进行探讨。然而，这些研究中的绝大多数仅仅分析了种植部门中的固碳措施。② 所测算的固碳潜力也因所采取的固碳措施的不同而有所差别。此外，这些研究缺乏关于相关减排成本的信息。需要注意的是，畜牧生产产生了大量的非二氧化碳温室气体，其所产生的甲烷和氧化亚氮分别占人类排放总量的44%和53%。③ 然而，以上研究都没有考虑中国畜牧生产排放的温室气体。为了更全面地理解种—畜复合经营农业部门中碳汇农业措施对于温室气体排放的影响，有必要对作物生产以及牲畜肠道发酵和粪便所产生的温室气体进行综合分析。在种—畜复合经营农业中，农业温室气体减排政策可能会引起种植部门、畜牧部门以及种—畜生产经营结构方面的变化。

考虑到中国近期在市场型碳减排机制方面的实践（如2012年开始设立的区域碳市场以及2017年开始谋划的全国性碳交易计划），政策制定者会乐于了解碳汇农业在温室气体减排方面的成本有效性。为了减轻实施碳交易计划方面的工作压力，相关监管部门不太愿意设立严格的减排目标。例如，天津碳市场的任务目标被设定为排放密度年均减少0.2%。作为一个相对的减排目标，降低排放密度并不直接意味着温室气体排放量的减少。与这些区域碳市场温和的减排目标相对应的是，市场上的碳价多在50~150元/tCO_2e波动。那么，一个与之相关的问题是，既然农业温室气体减排政策被设定在可比较的水平上，碳汇农业到底能够减少多少温室气体排放呢？这个问题的答案在

---

① Tang, K., Kragt, M. E., Hailu, A., Ma, C. Carbon farming economics: What have we learned? [J]. Journal of Environmental Management, 2016, 172: 49 - 57. Tang, K., Hailu, A., Kragt, M. E., Ma, C. The response of broadacre mixed crop-livestock farmers to agricultural greenhouse gas abatement incentives [J]. Agricultural Systems, 2018, 160: 11 - 20.

② Zhang, W. F., Dou, Z. X., He, P., et al. New technologies reduce greenhouse gas emissions from nitrogenous fertilizer in China [J]. Proceedings of the National Academy of Sciences, 2013, 110 (21): 8375 - 8380. Yuan, Z. Q., Yu, K. L., Epstein, H., et al. Effects of legume species introduction on vegetation and soil nutrient development on abandoned croplands in a semi-arid environment on the Loess Plateau, China [J]. Science of the Total Environment, 2016, 541: 692 - 700. Ji, Y., Ranjan, R., Burton, M. A bivariate probit analysis of factors affecting partial, complete and continued adoption of soil carbon sequestration technology in rural China [J]. Journal of Environmental Economics and Policy, 2017, 6 (2): 153 - 167.

③ IPCC. 2006 Guidelines for National Greenhouse Gas Inventories. Intergovernmental Panel on Climate Change [R]. National Greenhouse Gas Inventories Programme, IPCC, 2006.

很大程度上取决于农民对于减排政策的响应,在根本上取决于农民的边际减排成本。据笔者所知,目前还没有关于中国雨养农业区种—畜复合经营农民如何对减排政策进行响应,以及在种—畜复合经营农业中进行碳汇农业是否对于整体温室气体减排而言是一项具有成本有效性优势的选择的相关研究。本章将对这些问题进行深入探讨。

6.3 研究对象与研究方法

6.3.1 研究区域

本章的研究对象是位于中国北部黄土高原地区的种—畜复合经营农业。黄土高原位于黄河中上游地区,面积64万平方千米,范围为东经100.9~114.5度、北纬33.7~41.3度,是世界上黄土覆盖面积最大的地区,除少数石质山地外,高原上覆盖深厚的黄土层,厚度在50~80米之间,最厚达180米。黄土高原位于中国地貌的第二级阶梯及第二级阶梯向第三级阶梯过渡地带,从东部平原向西部山地过渡,西接日月山和乌鞘岭,东至太行山,南抵秦岭,北达阴山,地跨青海省西宁市以南,甘肃省河西走廊以东,宁夏回族自治区、内蒙古自治区包头市以南,陕西省秦岭以北,山西省以及河南省西部少数地区。黄土高原主要由台地、坡地以及高度侵蚀的山地所组成,海拔多在1000~1600米,平均海拔约为1200米。黄土高原地区户籍人口约为1.2亿,常住人口约为1.1亿,人口密度(167人/平方千米)比全国平均值高约23%。[1] 在中国,黄土高原地区可耕地面积仅次于黄淮海平原,位居全国第二。

黄土高原地区是中国最大的种—畜复合农业区。由于黄土高原的特殊地理位置,在漫长的历史时期中,一直处于种植畜牧两大系统的交互控制之中。种植畜牧系统的交互影响,使得种植畜牧两方面的技术都能深入到该区的农业系统中,也使得该区的种—畜复合农业系统结合较其他区域更好,动物生

---

[1] 宁夏回族自治区统计局:《宁夏统计年鉴2015》,中国统计出版社2015年版。

产更为突出,也建立了形成农业耦合系统的可能。① 目前该地区活羊供应量超过全国总量的20%,全区超过75%的土地被用于农业生产。② 虽然种—畜复合经营农业也存在于青藏高原,然而严酷的气候和自然环境使得青藏高原的农业生产规模要远小于黄土高原。

黄土高原地区属于典型的半干旱大陆性季风区,大陆性和季风不稳定性尤为突出,冬季受极地干冷气团影响,寒冷干燥,夏季受西太平洋副热带高压和印度洋低压影响,炎热湿润,气候四季分明。年均降水量自东南向西北快速递减。东南部如陕西省的关中平原,年均降水量可达600毫米。西部与北部如宁夏西北部地区以及鄂尔多斯高原西部地区,年均降水量仅为200毫米。黄土高原地区年均降水量均值约为450毫米,60%~80%的降水集中于7~8月。③ 该地区水热同季,光照充足,年日照时数2000~3000小时,年辐射总量502.416~669.888kJ/cm^2。年平均气温在8℃~14℃之间,年平均气温≥0℃的年积温3000℃~4000℃,无霜期120~200天,是中国光照资源充足的地区之一。④

郑景云等(2010)进一步将黄土高原分为干旱区、半干旱区和半湿润区三个气候区。其中,干旱区位于黄土高原西北部长城沿线以北及宁夏吴忠以北,年均气温7℃~12℃,年降雨量100~300毫米。气温年际、月际和日变化较大,降雨量少,风沙活动频繁,植被类型以草原和沙漠为主。半干旱区包括黄土高原多数地区,位于等降雨量线300毫米以东,陕西铜川以北及山西长治盆地以西,年均气温变幅较大,约为3.6℃~12℃,年降雨量300~600毫米,夏季风较弱,蒸发量大于降雨量,地貌以丘陵沟壑地貌为主。半湿润区面积相对较小,包括陕西铜川以南、山西南部及河南西部地区,年均气温8℃~15℃,年降雨量大于600毫米,夏季盛行东南风,该区地貌以河流阶地、平原和盆地为主。⑤

---

① 胥刚:《黄土高原农业结构变迁与农业系统战略构想》,兰州大学2015年博士学位论文。
② Liu, J., Liu, M., Zhuang, D., Zhang, Z., Deng, X. Study on spatial pattern of land-use change in China during 1995 - 2000 [J]. Science in China Series D: Earth Sciences, 2003, 46 (4): 373 - 384. 宁夏回族自治区统计局:《宁夏统计年鉴2015》,中国统计出版社2015年版。
③ Nolan, S., Unkovich, M., Yuying, S., Lingling, L., Bellotti, W. Farming systems of the Loess Plateau, Gansu Province, China [J]. Agriculture, Ecosystems & Environment, 2008, 124 (1): 13 - 23. 任婧宇、彭守璋、曹扬等:《1901~2014年黄土高原区域气候变化时空分布特征》,载《自然资源学报》2018年第4期。
④ 胥刚:《黄土高原农业结构变迁与农业系统战略构想》,兰州大学2015年博士学位论文。
⑤ 郑景云、尹云鹤、李炳元:《中国气候区划新方案》,载《地理学报》2010年第1期。

黄土高原的土地资源较为丰富。该区土地总量较大，全区共有黄绵土、褐土、钙土、黑垆土、风沙土等28个土类。黄绵土分布面积最广，粉砂含量通常在50%以上，含有大量的碳酸钙，易受降雨和径流侵蚀。[1] 黄土在被搬运沉积前，曾经历了基岩风化与原始成土过程，所以黄土既是成土母质也是具有一定肥力的土壤。黄土高原土壤富含碳酸钙和磷、钾、硼、锰等元素，有机质含量在历史上曾非常丰富，很适合农业生产的开展，这里也是中国农耕文明的重要发源地。然而，黄土高原是黄河的主要产沙源地，也是中国甚至世界上水土流失最严重的区域之一，土质疏松，暴雨冲刷强烈，长期缺乏植被保护，使得60%以上的土地存在着不同程度的水土流失，年平均泥沙流失量达到2000~2500t/km^2。[2] 水土流失使黄土高原丧失熟化土层，使当地土层的蓄水保湿能力降低，几近丧失耕种能力。随着数千年的开垦利用，目前黄土高原地区的土壤由于严重的土壤侵蚀和地表径流而肥力低下。当地农民施用氮、磷、钾肥来增加作物产量。改革开放前，农村普遍使用农家肥，还具有有机、缓释、促进微生物发育等优点，情况稍好；近年来随着城市化，农村种养结合的链条被打断，农家肥被弃用，大量使用化肥。由于土壤功能的缺失，化肥在黄土高原的效力非常低下，流失严重，进一步加剧了土壤的酸化等问题，形成恶性循环。当地土壤还具有低黏结力、高入渗率、保水能力较差的特征。[3]

从20世纪50年代末期开始，黄土高原地区随着人口增多，区域开发速度加快，人们毁林开荒、陡坡开垦耕地及过度放牧，使得地表植被覆盖迅速减少，水土流失进一步加剧，黄土高原生态环境进一步恶化。为改善这一现状，自20世纪80年代以来，该区开展了一系列水土保持工作，特别是1999年国家提出退耕还林还草工程，将坡度大于15°的耕地退耕为草地或林地，通过人工植被来恢复黄土高原植被生长，与此同时，封山育林，保护自然植

---

[1] 谢宝妮：《黄土高原近30年植被覆盖变化及其对气候变化的响应》，西北农林科技大学2016年博士学位论文。

[2] 田磊：《变化环境下黄土高原水文气候要素数值模拟及未来预测》，西北农林科技大学2019年博士学位论文。

[3] Wang, Y., Zhang, X., Huang, C. Spatial variability of soil total nitrogen and soil total phosphorus under different land uses in a small watershed on the Loess Plateau, China [J]. Geoderma, 2009, 150 (1-2): 141-149. 张凡、李长生：《气候变化影响的黄土高原农业土壤有机碳与碳排放》，载《第四纪研究》2010年第3期。

被恢复①，这使得黄土高原局部地区水土流失得到一定控制。随着植被覆盖度的增加，2000~2008年，黄土高原年径流量减少 61.8×10^8 立方米，土壤侵蚀减少 958 t/km²，② 生态环境得到明显改善。

为了进行生物经济学模拟分析，本章参照有关学者的研究结果（Messing et al.；Wang et al.）③，将黄土高原地区的土壤概括为四种类型（见表6-1）。

表6-1　　　　　　　　　　黄土高原的土壤类型

土壤类型	特征描述
砂土	砂质深度>150厘米；质地粗糙；SiO_2 含量>60%；有机质含量<0.15%
砂黄土	砂质深度>100厘米；质地粗糙；$CaCO_3$ 含量 7%~12%；有机质含量<0.3%
黄绵土	灰白疏松；粉砂壤质地；$CaCO_3$ 含量 10%~20%；有机质含量 0.5%~2%
基岩类	分散斑块状表土 20~40 厘米，底部基岩；粉砂岩和砂岩交替露头，厚度 5~10 米；pH>8

黄土高原地区农业属于典型的雨养种—畜复合农业。当地绝大多数的农户从事种—畜复合经营。④ 类似的农业系统还存在于中亚地区东南部、伊朗高原西部以及非洲南部的一些内陆地区。由于缺乏可利用的地表水，且地下水水位埋深多为50~80米，因而难以利用，当地农业生产主要依赖天然降水。⑤ 由于普遍存在水资源短缺，当地作物产量相对较低。例如，当地小麦产量平均每公顷为2.5~3.2吨，只相当于全国平均水平（每公顷5.7吨）的一半左右。⑥

当地农户每户农地面积多为0.5~3公顷（平均1.784公顷），农地土壤

---

① 田均良：《黄土高原生态建设环境效应研究》，气象出版社2010年版。

② Fu, W., Huang, M., Gallichand, J., Shao, M. Optimization of plant coverage in relation to water balance in the Loess Plateau of China [J]. Geoderma, 2012, 173 – 174: 134 – 144.

③ Messing, I., Chen, L., Hessel, R. Soil conditions in a small catchment on the Loess Plateau in China [J]. Catena, 2003, 54 (1 – 2): 45 – 58. Wang, Y., Zhang, X., Huang, C. Spatial variability of soil total nitrogen and soil total phosphorus under different land uses in a small watershed on the Loess Plateau, China [J]. Geoderma, 2009, 150 (1 – 2): 141 – 149.

④⑤ Wang, Y., Zhang, X., Huang, C. Spatial variability of soil total nitrogen and soil total phosphorus under different land uses in a small watershed on the Loess Plateau, China [J]. Geoderma, 2009, 150 (1 – 2): 141 – 149. 田均良：《黄土高原生态建设环境效应研究》，气象出版社2010年版。

⑥ Tsunekawa, A., Liu, G., Yamanaka, N., Du, S. Restoration and Development of the Degraded Loess Plateau, China [M]. Springer, Japan, 2014.

类型多样。① 通常有一多半的农地用于种植业，余下的用于畜牧业。然而，农地使用分配随着农民不同的偏好以及当地自然条件的差异（如生长季节降水、温度以及土壤类型）而有所不同。当地主要农产品包括谷类（如小麦和燕麦等）以及畜产品（如活羊）。

黄土高原农业活动历史悠久，种植的作物种类多样。当地的作物品种通常耐旱、耐寒、耐热，能够在贫瘠的土地生长。② 小麦是当地主要的作物，产量占作物总产量的30%，种植面积占耕地总面积的20%。③ 绝大多数的小麦收获后供农户家庭使用。传统上，小麦与其他豆科作物（如豌豆）、燕麦或玉米进行轮作。从20世纪70年代末以来，当地开始大规模种植经济作物（如油菜）。

许多当地农户也种植牧草，与作物进行轮作。苜蓿是当地普遍种植的一种多年生深根豆科牧草，在当地的种植历史超过2000年。近年来，苜蓿被广泛地种植于黄土高原地区，用以发展畜牧生产以及防止水土流失。④ 豆科牧草还能够通过生物固氮作用增加土壤中的氮素含量，从而提升后茬作物的产量。其他牧草品种包括小冠花（coronilla varia）、红豆草（onobrychis viciifolia）、冰草（agropyron cristatum）以及鸭茅（dactylis glomerata）等。牧草生物量的数量主要受到天气条件、土壤种类、放牧压力和肥料等因素的影响。牧草的生产通常始于春季初雨时；在夏末或秋初时牧草生物量达到顶峰；在秋末枯萎，但仍可作为冬季牲畜的饲料。

该地区绝大多数的农户都饲养绵羊。当地绝大部分的绵羊种群保持自然繁育状态。牧放的羊群用作生产羊肉、羊毛或羊绒以供出售。除了牧草，绵羊的饲料还包括作物秸秆和干草。有时农民会用谷物作为维持羊群营养的补充饲料。近年来，豌豆以及油菜也被用作干旱季节的饲料。

---

① 宁夏回族自治区统计局：《宁夏统计年鉴2015》，中国统计出版社2015年版。延安市统计局：《延安统计年鉴2016》，延安市统计局，2017。

② Tsunekawa, A., Liu, G., Yamanaka, N., Du, S. Restoration and Development of the Degraded Loess Plateau, China [M]. Springer, Japan, 2014.

③ Nolan, S., Unkovich, M., Yuying, S., Lingling, L., Bellotti, W. Farming systems of the Loess Plateau, Gansu Province, China [J]. Agriculture, Ecosystems & Environment, 2008, 124 (1): 13–23.

④ Yuan, Z. Q., Yu, K. L., Epstein, H., et al. Effects of legume species introduction on vegetation and soil nutrient development on abandoned croplands in a semi-arid environment on the Loess Plateau, China [J]. Science of the Total Environment, 2016, 541: 692–700.

6.3.2 全农场生物经济模型

本章的分析主要利用唐凯等所构建的卡坦宁全农场生物经济模型进行。[①] 模型的初始版本（Hailu et al.）是一个旨在最大化毛利的动态数学规划模型。[②] 其中，毛利等于农场收入扣除变动成本以及间接费用。模型基于农地管理单元利用历史，对农地利用序列选择进行优化。模型考虑了半干旱雨养种—畜复合农业在生物、技术、管理以及财务约束等方面的特征。模型中涵盖了几百种农业活动，包括每一个农地管理单元上的作物—牧草轮作、不同类别牲畜的饲料供给与使用、牲畜的繁殖等。模型的最优解描述的是基于农场可利用资源，在一系列资源、逻辑以及技术等条件的约束下，能够产生最大化毛利润的一系列农场生产活动的集合。

为了探究农业减排政策对于农地利用、毛利以及温室气体排放的影响，本书为最新版的模型增加了测算农业温室气体年排放量的模块。所使用的农业温室气体测算方法来自政府间气候变化专门委员会所公布的清单方法[③]，并且对相应参数进行了调整以反映当地的半干旱农业系统特征[④]（参见附录）。所有的温室气体排放量转化为 CO_2e。本章考虑了四类农业排放源，包括作物秸秆、化肥施用、牲畜以及固氮植株。温室气体减排量依据农业减排政策所引起农业活动的变化来测算。

本章所考虑的农业温室气体减排政策是一种农业碳税，要求农民为其在生产中所排放的每一吨温室气体缴纳一定金额的税。此项碳税不包括免税排放额，且税率固定，不随着温室气体排放量的变化而浮动。碳税的税率从 0 上浮到 500 元/tCO_2e，每次上浮的幅度为 50 元/tCO_2e。之所以选择固定税率

[①] Tang, K., Hailu, A., Kragt, M. E., Ma, C. The response of broadacre mixed crop-livestock farmers to agricultural greenhouse gas abatement incentives [J]. Agricultural Systems, 2018, 160: 11 – 20.

[②] Hailu, A., Durkin, J., Sadler, R., Nordblom, T. L. Agent-based modelling study of shadow, saline water table management in the Katanning catchment, Western Australia [R]. Research Report for RIRDC Project No. PRJ – 000578, 2011.

[③] 具体可参见 https://www.ipcc-nggip.iges.or.jp/public/2006gl/vol4.html.

[④] Tang, K., Hailu, A., Kragt, M. E., Ma, C. Marginal abatement costs of greenhouse gas emissions: broadacre farming in the Great Southern Region of Western Australia [J]. Australian Journal of Agricultural and Resource Economics, 2016, 60: 459 – 475. Tang, K., Hailu, A., Kragt, M. E., Ma, C. The response of broadacre mixed crop-livestock farmers to agricultural greenhouse gas abatement incentives [J]. Agricultural Systems, 2018, 160: 11 – 20.

碳税作为农业温室气体减排政策，主要是考虑分析上的便利性。然而，该全农场生物经济模型还可与其他类型农业温室气体减排政策结合使用，如基于配额拍卖的减排政策。

模型模拟的是黄土高原地区的一个典型农户。农户耕地面积1.784公顷，包括12个农地管理单元。为了反映当地典型土壤条件，①典型农户的大部分耕地土壤设定为砂黄土和黄绵土。②农户从事种—畜复合经营。模型所考虑的作物包括小麦、燕麦、油菜以及豌豆。作物与牧草进行轮作。农户饲养的代表性牲畜是绵羊。绵羊饲料相关参数参见唐凯等。③年均降水量设定为445毫米，以反映当地半干旱大陆性季风气候的降水水平。④

作物与畜牧商品的价格是收购价，以2015年人民币表示。模型假设种—畜复合经营农民依据作物与牲畜的预期市场价格、农业生产成本以及当地的环境条件（如土壤类型、生长季节降雨量和作物轮作效应）来做出农业生产决策。

6.4 研究结果

本书先给出的是没有农业碳税时的模拟结果。接下来展示的是不同农业碳税税率情景下的模拟结果。本章的模拟周期是10年。所展示结果均为10年周期的均值。

6.4.1 零农业碳税情景

当农业碳税税率为0时，通过优化农场的种—畜生产经营所得到的最大

---

① Messing, I., Chen, L., Hessel, R. Soil conditions in a small catchment on the Loess Plateau in China [J]. Catena, 2003, 54 (1-2): 45-58. Wang, Y., Zhang, X., Huang, C. Spatial variability of soil total nitrogen and soil total phosphorus under different land uses in a small watershed on the Loess Plateau, China [J]. Geoderma, 2009, 150 (1-2): 141-149.

② 作者就相关参数设定与来自中国科学院地理科学与资源研究所、华中农业大学以及西北农林科技大学相关领域的有关专家进行了探讨。专家对所设定的参数表示了肯定。

③ Tang, K., Hailu, A., Kragt, M. E., Ma, C. The response of broadacre mixed crop-livestock farmers to agricultural greenhouse gas abatement incentives [J]. Agricultural Systems, 2018, 160: 11-20.

④ Nolan, S., Unkovich, M., Yuying, S., Lingling, L., Bellotti, W. Farming systems of the Loess Plateau, Gansu Province, China [J]. Agriculture, Ecosystems & Environment, 2008, 124 (1): 13-23.

毛利年均为每公顷 3540.6 元（见图 6-1）。约有一半的农地用作种植作物，余下农地用作种植牧草供羊群食用。平均而言，优化后的种—畜生产经营结构中，约有 49.4%（0.882 公顷）、18.5%（0.329 公顷）、21.1%（0.376 公顷）、10.75%（0.192 公顷）和 0.3%（0.005 公顷）的农地被分别用作种植牧草、小麦、燕麦、油菜和豌豆。主要的作物轮作组合包括持续性牧草、牧草—燕麦组合、牧草—小麦—燕麦组合以及牧草—油菜—小麦组合（见表 6-2）。需要注意的是，在该模型所包含的 12 个农地管理单元中，各单元所选择的种—畜生产活动因为土壤类型的不同而各异。这些种—畜生产活动从总体上保证了整个农业生产系统的毛利最大化。作物的产量约为每公顷 3 吨。以上各模拟结果与已有的关于黄土高原地区半干旱农业系统的研究结论基本一致。[①]

在零农业碳税的优化情景中，农户温室气体年均排放总量为 4.06 tCO_2e，折合年均每公顷 1.95 tCO_2e。研究周期内平均而言，畜牧生产排放了占农户总排放量的 85% 的温室气体（年均 3.74 tCO_2e）。固氮类作物是第二大排放源，其排放比重约占农户总排放量的 12%。化肥施用以及作物秸秆所产生的排放则相对较少。以上结果与相似半干旱复合农业系统相关研究的结果一致。[②]

6.4.2 正农业碳税情景

表 6-2 展示的是不同农业碳税情景下毛利最大化时的轮作组合。图 6-1 展示的是相应的农户温室气体排放量以及优化后的毛利。图 6-2 展示的是不同农业碳税情景下优化后的农地利用分配。图 6-3 比较了农业碳税对于总排放量以及牲畜排放量的影响。

[①] Tsunekawa, A., Liu, G., Yamanaka, N., Du, S. Restoration and Development of the Degraded Loess Plateau, China [M]. Springer, Japan, 2014.

[②] Thamo, T., Kingwell, R. S., Pannell, D. J. Measurement of greenhouse gas emissions from agriculture: Economic implications for policy and agricultural producers [J]. Australian Journal of Agricultural and Resource Economics, 2013, 57: 234-252. Tang, K., Hailu, A., Kragt, M. E., Ma, C. Marginal abatement costs of greenhouse gas emissions: broadacre farming in the Great Southern Region of Western Australia [J]. Australian Journal of Agricultural and Resource Economics, 2016, 60: 459-475.

表 6-2　　不同农业碳税情景下毛利最大化时的轮作组合

农业碳税税率（元/tCO$_2$e）	轮作组合			
	砂土	砂黄土	黄绵土	基岩类
0	POO, PRO, PPOD	POO, POP, POW, PRO, PWO, PRW, WPRW	PPP, POP, PPO, PWP, PPW, PRW	PRW, PRWP
50	POW, PWW, PRW, WPWO	PPP, POO, POP, PRW, WPRW	POP, POW, PWO, PWW, PWPW, PDW, PWD, PRW, DOP, OODR	POO, POOD
100	PPP, PRW, PDR	PWW, PPW, PRW, PDW, WPRW	PWW, WPW, WPP, PRW, PDW, PPD, WPD	PPP, WDP
150	PPO, WPWO, PPO	PRW, PDW, PWD, DRW	PWW, PDW, PWD, PDWD	PWW, PRW, PRWD
200	PWW, PDW, DPDW	PWW, PRW, PDW, WPDW	PWW, PRW, PDW, PWD, PDWD, DRW, DPDW	PPO, PDW
250	PRW, PDR, DRW	PWW, PRW, PDW, PWD	PWW, PRW, WPRW, PDW, PWD, PDWD	PWW, PDW, PDWD
300	PWW, PWD, PDW	PWW, PDW, PDWD	PWW, PDW, PWD, PDR, WPDO	PWW, DRW
350	PWW, PDW, DPDW, WPR	PWW, PDW, PWD, DPDW	PWW, PDW, PWD, DPDW, DRW	DRW, DRWD
400	PDW, PWD, DRW, DCWR	PDW, PDWD, DRW	PWW, PDW, PWD, PDWD, DPDW	PDR, DRW
450	PDW, PWD, PDR, DRW	PWW, PDW, PWD, PDWD, DRW	PWW, PDW, PWD, PDWD, DPDW, RDR	PDW, PDWD
500	PDW, PDWD, WDR	PWW, PDW, PWD, PDWD, DPDW, WDR	PWW, PDW, PWD, PDWD, DPDW, DRD	PWW, PDW, PDWD

注：P = 牧草；W = 小麦（triticum aestivum）；O = 燕麦（avena sativa）；R = 油菜（brassica napus）；D = 豌豆（pisum sativum）。

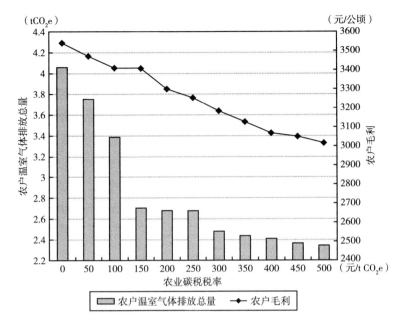

图6-1 不同农业碳税情景下优化后的农户温室气体总排放量与毛利

当农业碳税分别为50元/tCO_2e、100元/tCO_2e和150元/tCO_2e时,年均最大毛利分别下降到每公顷3471.2元、每公顷3409.3元和每公顷3407.4元;农民种植更多的小麦和豌豆,减少了牧草、燕麦以及油菜的种植面积(见图6-2);更多包含豌豆的轮作组合被包含到最优生产经营方案中(见表6-2)。当农业碳税税率高于100元/tCO_2e时,主要的轮作组合是牧草—小麦组合和牧草—豌豆—小麦组合,而包含燕麦的轮作组合没有被纳入最优生产经营方案中(见表6-2)。当农业碳税分别为50元/tCO_2e、100元/tCO_2e和150元/tCO_2e时,农户温室气体排放总量与零碳税情景相比分别减少7.6%、16.6%和33.3%。畜牧生产和作物生产所减少的温室气体排放下降趋势相近。在牧草—豌豆—小麦轮作组合中,作为豌豆后茬作物的小麦的产量为每公顷3.6吨,比其他轮作组合中小麦的产量高11%。

随着农业碳税税率的进一步提高(150~500元/tCO_2e),优化后的毛利呈下降态势。然而,牧草种植面积以及温室气体排放总量的减少量均较为有限。值得注意的是,在所有情景中,小麦和豌豆都是主要的作物。此外,牧草—小麦与牧草—豌豆—小麦轮作组合是所有类型土壤的主要轮作组合。当碳税税率高达500元/tCO_2e时,农户温室气体排放总量下降到2.34tCO_2e,

比零碳税情景低42.2%。优化后的毛利为每公顷3015.4元。此时，42.9%的农地被用作种植豌豆，余下的农地大致被平分，用于种植小麦和牧草。

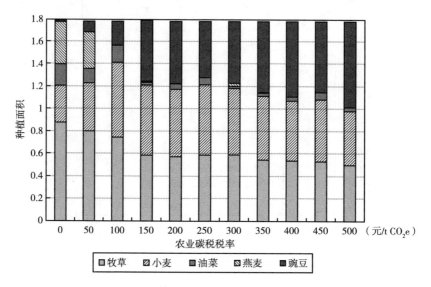

图6-2　不同农业碳税情景下优化后的农地利用分配

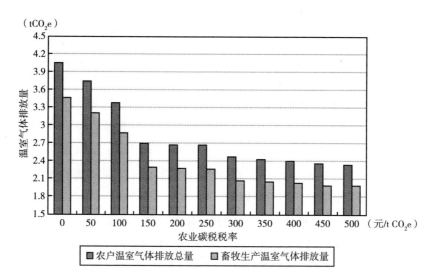

图6-3　不同农业碳税情景下农户温室气体排放总量
与牲畜温室气体排放量

6.5 进一步讨论

分析结果显示，随着农业碳税税率的提高，黄土高原地区种—畜复合经营农业中作物种植的主导程度会不断提升。已有研究显示[①]，畜牧生产是农业温室气体的主要排放源，畜牧生产与作物种植相比，前者的排放密度要远高于后者。本章分析结果显示，畜牧生产的年均温室气体排放量约为每公顷 3.85 tCO_2e，而作物种植排放量小于每公顷 0.65 tCO_2e。当农业温室气体排放面临碳税的规制时，作物种植与畜牧生产相比，碳税给前者所带来的外部成本要远小于给后者。因此，农民出于对碳税的理性响应，会选择限制包括牲畜养殖和种植牧草在内的高排放密度畜牧生产，转而扩张作物种植规模。

随着中国居民膳食结构中动物性食物的增加，[②] 温室气体减排政策（如碳税）所引起畜牧生产的缩减，可能会导致市场上出现较大的供给缺口。对于这一矛盾，一项基础的解决方案是进一步提升畜牧生产的生产力和牲畜健康水平。改善牲畜的遗传潜力，提升牲畜的繁殖率、健康比例以及活体增重率，是降低每单位产品温室气体排放量的有效办法。[③] 包括使用饲料添加剂、改善饲料可消化性以及妥善处理牲畜粪便在内的技术与管理干预，也有助于实现增加畜牧产品供给与减少农业温室气体排放之间的平衡。供给缺口也为包括澳大利亚、新西兰、美国以及巴西在内的主要国际供货商

[①] Fiala, N. Meeting the demand: An estimation of potential future greenhouse gas emissions from meat production [J]. Ecological Economics, 2008, 67 (3): 412 - 419. Thamo, T., Kingwell, R. S., Pannell, D. J. Measurement of greenhouse gas emissions from agriculture: Economic implications for policy and agricultural producers [J]. Australian Journal of Agricultural and Resource Economics, 2013, 57: 234 - 252. Herrero, M., Henderson, B., Havlík, P., et al. Greenhouse gas mitigation potentials in the livestock sector [J]. Nature Climate Change, 2016, 6 (5): 452 - 461. Tang, K., Hailu, A., Kragt, M. E., Ma, C. The response of broadacre mixed crop-livestock farmers to agricultural greenhouse gas abatement incentives [J]. Agricultural Systems, 2018, 160: 11 - 20.

[②] Hawkins, J., Ma, C., Schilizzi, S., Zhang, F. China's changing diet and its impacts on greenhouse gas emissions: an index decomposition analysis [J]. Australian Journal of Agricultural and Resource Economics, 2018, 62 (1): 45 - 64.

[③] Herrero, M., Henderson, B., Havlík, P., et al. Greenhouse gas mitigation potentials in the livestock sector [J]. Nature Climate Change, 2016, 6 (5): 452 - 461.

创造了机会。

研究结果还显示,当农业碳税税率不断提高时,农民倾向于在种植作物时减少包含油菜和燕麦的轮作组合,而更多地选择包含豌豆的轮作组合。这是因为作物所产生的温室气体,由其在生产过程中所需施用氮肥的数量,以及一些作物器官(如根、秸秆)的氮浓度所决定。① 已有研究表明,在半干旱农业系统中,包含燕麦和油菜的轮作组合与包含豌豆的轮作组合相比,后者在生长过程中需要更少的氮肥,因而会排放更少的温室气体。② 豌豆根部共生的豌豆根瘤菌具有固氮作用,其固定的氮素能满足豌豆生长期部分所需量。因此,种植豌豆不需要大量的氮肥,从而减少相关的温室气体排放。此外,豌豆能够通过多种途径提升后茬作物的产量。在牧草—豌豆—谷物轮作组合中,谷物作为豌豆的后茬作物,其产量能够得到提升。例如,本书研究发现,牧草—豌豆—小麦与其他轮作组合相比,前者小麦的产量要比后者增加11%。如果增加轮作中豌豆的种植频率,降低土壤中氮残留量过高对于固氮的负面影响,以及提高来自豌豆茬中氮矿化与后作氮需求高峰期之间的同步性,后作的产量可以得到进一步提高。③ 因此,农民在得到农业温室气体减排政策措施激励的情况下,会倾向于将减少包含燕麦和油菜的轮作组合,增加包含豌豆的轮作组合。

更重要的是,本章研究发现,农业碳税税率相对较小的提升可以产生显著的农户温室气体减排效果,减排所需成本较低。当农业碳税税率分别为50元/tCO_2e、100元/tCO_2e和150元/tCO_2e时,农户温室气体排放总量分别减少7.6%、16.6%和33.3%。相应的毛利损失仅分别为2%、3.7%和3.8%。

以上结果说明,减少7.6%、16.6%和33.3%农户温室气体排放总量的边际减排成本要分别低于50元/tCO_2e、100元/tCO_2e和150元/tCO_2e。已有文献给出了中国农业部门"自下而上"的边际减排成本曲线,结果显示,减

---

① Gan, Y. T., Campbell, C. A., Jansen, H. H., et al. Carbon input to soil by oilseed and pulse crops in semiarid environment [J]. Agriculture, Ecosystem & Environment, 2009, 132: 290–297.

② Kirkegaard, J., Christen, O., Krupinsky, J., Layzell, D. Break crop benefits in temperate wheat production [J]. Field Crops Research, 2008, 107 (3): 185–195. Rajaniemi, M., Mikkola, H., Ahokas, J. Greenhouse gas emissions from oats, barley, wheat and rye production [J]. Agronomy Research, 2011, 9 (1): 189–195.

③ Herridge, D. F., Peoples, M. B., Boddey, R. M. Global inputs of biological nitrogen fixation in agricultural systems [J]. Plant and Soil, 2008, 311 (1–2): 1–18. Tang, K., Hailu, A., Kragt, M. E., Ma, C. The response of broadacre mixed crop-livestock farmers to agricultural greenhouse gas abatement incentives [J]. Agricultural Systems, 2018, 160: 11–20.

少 40% 排放量的边际减排成本低于 123 元/tCO₂e。[①] 本章的结果与其研究结果大致一致。

考虑到中国近期试行的市场化碳减排的经验，本章结果可能对设计全国性碳交易计划有重要启示。在区域碳市场中，温和的减排目标所对应的碳价已经达到 50~150 元/tCO₂e。本章结果显示，在农业温室气体减排政策被设定在可比较水平的情况下，可以减少大量的温室气体排放。在中国雨养种—畜复合农业中进行碳汇农业，是一项极具成本有效性优势的选择。

在解读本章所得出的结论时需注意以下问题。

第一，本章农业温室气体排放量的估计建立在有关农产品价格、土壤类型以及气候条件的一系列假设基础之上。本书的结果是基于处于半干旱气候条件下、以雨养种—畜复合农业为代表的中国黄土高原地区。研究结果也许能够用来理解在中亚东南部地区、伊朗高原西部以及南部非洲一些内陆地区类似农业系统中的农业减排。然而，中国以及一些国家其他的农业区域在农业系统、土壤类型和气候上的差异，以及农产品市场价格的波动，都有可能对农民在农地利用、生产毛利以及温室气体排放方面的响应产生影响。

第二，本章研究没有考虑由农业减排活动所带来的潜在的协同效益，如生态多样性的保护以及土壤品质的改良。[②] 黄土高原地区存在着严重的土壤侵蚀和地表径流，土壤通常肥力较低。[③] 通过采取如轮作和保护性耕作在内的碳汇农业措施，农民可以减少土壤侵蚀、改进土壤肥力。类似私人的协同效应能够进一步地降低温室气体排放的边际减排成本。此外，碳汇农业也能够带来其他的公共效用，如生态多样性保护等。对于这些私人和公共协同效应的综合分析可能说明，政府应当提供更优厚的农业温室气体减排政策。

第三，应注意，采取碳汇农业措施会产生一系列与农业生产非直接相关

---

① Wang, W., Koslowski, F., Nayak, D. R., et al. Greenhouse gas mitigation in Chinese agriculture: distinguishing technical and economic potentials [J]. Global Environmental Change, 2014, 26: 53 – 62.

② Tang, K., Kragt, M. E., Hailu, A., Ma, C. Carbon farming economics: What have we learned? [J]. Journal of Environmental Management, 2016, 172: 49 – 57.

③ Wang, Y., Zhang, X., Huang, C. Spatial variability of soil total nitrogen and soil total phosphorus under different land uses in a small watershed on the Loess Plateau, China [J]. Geoderma, 2009, 150 (1 – 2): 141 – 149.

的额外成本，如交易成本和学习成本等。① 本章所使用的模型没有考虑这些额外成本。然而在实际操作中，这些额外成本可能给中国经营种—畜复合农业农民采取碳汇农业措施带来一定的阻碍。

6.6 本章小结

本章利用一个全农场生物经济模型，分析了在程度不同的以农业碳税为代表的农业温室气体减排政策情景下，中国黄土高原地区半干旱雨养种—畜复合农业系统在农地使用、农业生产行为以及农户温室气体排放量方面的变化。

研究结果显示，由于牲畜是主要的农业温室气体排放源，优化后的复合农业系统中种植业的比例将上升，以减少农业生产所排放的温室气体。受农业碳税的影响，在优化后的生产经营组合中，农民倾向于包含更多的基于豌豆的轮作组合，而减少基于燕麦以及油菜的轮作组合。此外，碳税税率相对较小的提升可以产生较大的农业温室气体减排效果，且减排成本相对较小。分析结果表明，黄土高原地区的种—畜复合经营农民减少 16.6% 和 33% 温室气体排放的边际减排成本，以 2015 年人民币计算不高于 100 元/tCO_2e 和 150 元/tCO_2e。

考虑到中国区域碳市场的实际碳价，本章结果表明，在中国雨养种—畜复合农业中减少温室气体排放，是一项相对低成本的选择。为中国半干旱雨养种—畜复合农业部门提供农业温室气体减排政策，不仅能够改变农户农地利用和农业生产行为，显著减少农业温室气体排放，还可以以极具成本有效性优势的方式，实现全社会的有效减排。

---

① Bakam, I., Balana, B. B., Matthews, R. Cost-effectiveness analysis of policy instruments for greenhouse gas emission mitigation in the agricultural sector [J]. Journal of Environmental Management, 2012, 112: 33-44. Tang, K., Kragt, M. E., Hailu, A., Ma, C. Carbon farming economics: What have we learned? [J]. Journal of Environmental Management, 2016, 172: 49-57.

第7章 黄土高原地区雨养农业小农户对气候变化影响的适应

7.1 引　　言

在过去几十年，气候变化对农业部门产生了广泛的负面影响。① 这些负面影响严重与否，往往取决于农业生产通过适应气候变化来减缓气候变化负面影响的程度。② 在许多发展中国家，上述依赖关系表现得尤其突出。当地小农户③由于缺乏必要的适应能力，在面对气候变化的负面影响时变得更加脆弱。④ 因此，考虑可能的适应措施，避免过高估计气候变化对小农生产的影响，是极其重要的。然而，绝大多数关注气候变化对发展中国家农业生产

---

① Arslan, A., Belotti, F., Lipper, L. Smallholder productivity and weather shocks: Adoption and impact of widely promoted agricultural practices in Tanzania [J]. Food Policy, 2017, 69: 68–81. FAO. FAO Strategy on Climate Change [R]. FAO, Rome, 2017. World Bank. Looking beyond the horizon: how climate change impacts and adaptation responses will reshape agriculture in Eastern Europe and Central Asia [R]. World Bank, Washington, D. C., 2013.

② Thamo, T., Addai, D., Pannell, D. J., et al. Climate change impacts and farm-level adaptation: Economic analysis of a mixed cropping-livestock system [J]. Agricultural Systems, 2017, 150: 99–108.

③ 据联合国粮食及农业组织统计，发展中国家的小农家庭数量约为4.75亿。参见：联合国粮食与农业组织：《2016粮食及农业状况：气候变化、农业与粮食安全》，2016年。

④ Salazar-Espinoza, C., Jones, S., Tarp, F. Weather shocks and cropland decisions in rural Mozambique [J]. Food Policy, 2015, 53: 9–21. Trinh, T. Q., Rañola Jr, R. F., Camacho, L. D., Simelton, E. Determinants of farmers' adaptation to climate change in agricultural production in the central region of Vietnam [J]. Land Use Policy, 2018, 70: 224–231.

影响的研究，并没有将农业对气候变化的适应考虑在内。①

本章分析了气候变化对中国北部黄土高原地区的影响，并考虑了相关适应情景。黄土高原地区是东亚最重要的雨养农业区域，当地农业由于对气候尤其是可利用降水高度敏感而十分脆弱。② 1956～2011 年，虽然年际降水量存在波动，但是黄土高原地区年均降水量大体上减少了 10%，③④ 且降水减少的趋势预计还将持续。⑤ 此外，中国已设定了专门的农业温室气体减排目标，以减缓气候变化的影响，具体包括到 2020 年实现农田氧化亚氮排放达到峰值，以及单位 GDP 碳排放减少 18%。为了实现这些减排目标，预计政府将出台相应的财政支持措施以及其他政策。作为一种潜在的财政手段，中国可能在未来对农业部门所排放的温室气体征收排放税，而黄土高原地区雨养复合经营农业也可能被其所涵盖。⑥

需要指出的是，据作者所知，迄今为止还没有国家直接对农业所排放

---

① Birthal, P. S., Negi, D. S., Khan, M. T., Agarwal, S. Is Indian agriculture becoming resilient to droughts? Evidence from rice production systems [J]. Food Policy, 2015, 56: 1 – 12. Chen, Z. M., Ohshita, S., Lenzen, M., et al. Consumption-based greenhouse gas emissions accounting with capital stock change highlights dynamics of fast-developing countries [J]. Nature Communications, 2018, 9 (1): 1 – 9. Guan, K., Sultan, B., Biasutti, M., Baron, C., Lobell, D. B. Assessing climate adaptation options and uncertainties for cereal systems in West Africa [J]. Agricultural and Forest Meteorology, 2017, 232: 291 – 305. van Valkengoed, A. M., Steg, L. Meta-analyses of factors motivating climate change adaptation behaviour [J]. Nature Climate Change, 2019, 9 (2): 158 – 163.

② Fu, B., Liu, Y., Lü, Y., et al. Assessing the soil erosion control service of ecosystems change in the Loess Plateau of China [J]. Ecological Complexity, 2011, 8 (4): 284 – 293. Li, X., Philp, J., Cremades, R., Roberts, A., Liang, H., Li, L., Yu, Q. Agricultural vulnerability over the Chinese Loess Plateau in response to climate change: exposure, sensitivity, and adaptive capacity [J]. Ambio, 2016, 45 (3): 350 – 360. Wang, S., Fu, B., Piao, S., et al. Reduced sediment transport in the Yellow River due to anthropogenic changes [J]. Nature Geoscience, 2016, 9 (1): 38. Wang, S., Fu, B., Chen, H., Liu, Y. Regional development boundary of China's Loess Plateau: Water limit and land shortage [J]. Land Use Policy, 2018, 74: 130 – 136.

③ 在陕西北部、山西中北部以及汾河流域，年均降水量减少的趋势更为明显。

④ Xin, Z., Yu, X., Li, Q., Lu, X. X. Spatiotemporal variation in rainfall erosivity on the Chinese Loess Plateau during the period 1956 – 2008 [J]. Regional Environmental Change, 2011, 11 (1): 149 – 159. 赵一飞、邹欣庆、张勃等：《黄土高原甘肃区降水变化与气候指数关系》，载《地理科学》2015 年第 10 期。程楠楠、何洪鸣、逯亚杰等：《黄土高原近 52 年降水时空动态特征》，载《山东农业大学学报（自然科学版）》2016 年第 3 期。

⑤ Huang, J., Yu, H., Guan, X., Wang, G., Guo, R. Accelerated dryland expansion under climate change [J]. Nature Climate Change, 2016, 6 (2): 166 – 171.

⑥ Liu, L. C., Wu, G. The effects of carbon dioxide, methane and nitrous oxide emission taxes: An empirical study in China [J]. Journal of Cleaner Production, 2017, 142: 1044 – 1054.

的温室气体征税。然而,基于生命周期的视角,农业温室气体在那些建立了碳排放交易市场的地区已经被间接地征税,例如,在中国启动了碳排放权交易试点的7个地区。此外,农业碳税正在被越来越多的学者和政策制定者所关注。[①] 包括中国与一些欧洲国家的许多研究者已经意识到,农业碳税是一种能有效减少温室气体排放的潜在市场型政策工具。考虑到中国强力推进的温室气体减排国家目标,农业部门所排放的温室气体存在着被直接征税的可能。

本章的目的在于分析黄土高原地区在未来气候变化条件下的农业生产以及利润。在现有研究的基础上,本章进一步分析了在发展中国家采取适应措施的情况下,降水减少和农业碳税征收的综合作用,以及小农户对这些综合作用的适应行为。具体而言,本章将分析以下几个问题。第一,在一系列潜在的自然—社会情景中,气候变化对农业利润有何影响;第二,在采取了适应措施的情况下,农业温室气体排放量如何变化;第三,基于现有可利用的适应措施,小农户如何能够适应气候变化。

7.2 相关研究述评

一些研究关注了发展中国家小农农业,分析了降水减少所造成的农业产出以及利润的变化。[②] 这些研究多使用具体的作物或畜牧模型来进行仿真。在实际中,小农会采取可利用的适应措施,以减少由产出变化以及如农业碳税在内的负激励所造成的潜在经济损失。然而,鲜有研究分析了在发展中国家小农户采取适应措施情况下,降水减少以及农业碳税征收的综合作用。

---

① Tang, K., Kragt, M. E., Hailu, A., Ma, C. Carbon farming economics: What have we learned? [J]. Journal of Environmental Management, 2016, 172: 49–57.

② Li, Z., Liu, W., Zhang, X., Zheng, F. Assessing the site-specific impacts of climate change on hydrology, soil erosion and crop yields in the Loess Plateau of China [J]. Climatic Change, 2011, 105 (1–2): 223–242. Chen, H., Zhao, Y., Feng, H., Li, H., Sun, B. Assessment of climate change impacts on soil organic carbon and crop yield based on long-term fertilization applications in Loess Plateau, China [J]. Plant Soil, 2015, 390 (1–2): 401–417. Trinh, T. Q., Rañola Jr, R. F., Camacho, L. D., Simelton, E. Determinants of farmers' adaptation to climate change in agricultural production in the central region of Vietnam [J]. Land Use Policy, 2018, 70: 224–231.

研究所涵盖适应措施的综合作用主要有两种方法。第一种方法需要研究者对潜在适应措施进行组合，然后分析每一种所选组合的影响。① 研究者可以完全决定哪些适应组合被选择，以确保能够对特定适应措施进行有效分析。然而在复杂的复合经营农业地区（如黄土高原地区），盲目选择适应组合是不合理的。由于难以对所有可能的措施进行充分比较，研究者可能会忽视一些潜在有效的适应组合。② 第二种方法是对所有可能的适应组合进行自动与系统的优化分析。③ 该方法不需要识别最优适应组合。在具有众多自然—社会参数的复杂系统环境下，该方法相较第一种方法更为有效。近年来，该方法被运用于有关发达国家大农场的分析中。④ 然而，鲜有学者利用该方法分析发展中国家小农户应对气候变化综合作用的适应组合。

　　现有关于黄土高原地区气候变化作用的研究多关注单一农业生产部门。⑤ 总体上，这些研究通过分析其所关注的农业生产部门，认为气候变化对当地的农业系统有着显著影响。然而，这些研究往往忽略了当地农业系统中众多组成部分的综合变化。事实上，在黄土高原地区，当地大部分农户从事如种

---

① Chalise, S., Naranpanawa, A. Climate change adaptation in agriculture: A computable general equilibrium analysis of land-use change in Nepal [J]. Land Use Policy, 2016, 59: 241 – 250. Fahad, S., Wang, J. Farmers' risk perception, vulnerability, and adaptation to climate change in rural Pakistan [J]. Land Use Policy, 2018, 79: 301 – 309.

② Challinor, A. J., Watson, J., Lobell, D. B., Howden, S. M., Smith, D. R., Chhetri, N. A meta-analysis of crop yield under climate change and adaptation [J]. Nature Climate Change, 2014, 4 (4): 287 – 291.

③ Thamo, T., Addai, D., Pannell, D. J., et al. Climate change impacts and farm-level adaptation: Economic analysis of a mixed cropping-livestock system [J]. Agricultural Systems, 2017, 150: 99 – 108.

④ Farquharson, R., Abadi, A., Finlayson, J., et al. EverFarm® – Climate adapted perennial-based farming systems for dryland agriculture in southern Australia [R]. National Climate Change Adaptation Research Facility, Gold Coast Australia, 2013. Thamo, T., Addai, D., Pannell, D. J., et al. Climate change impacts and farm-level adaptation: Economic analysis of a mixed cropping-livestock system [J]. Agricultural Systems, 2017, 150: 99 – 108.

⑤ Li, Z., Liu, W., Zhang, X., Zheng, F. Assessing the site-specific impacts of climate change on hydrology, soil erosion and crop yields in the Loess Plateau of China [J]. Climatic Change, 2011, 105 (1 – 2): 223 – 242. Liu, W., Sang, T. Potential productivity of the Miscanthus energy crop in the Loess Plateau of China under climate change [J]. Environmental Research Letters, 2013, 8 (4), 044003. Chen, H., Zhao, Y., Feng, H., Li, H., Sun, B. Assessment of climate change impacts on soil organic carbon and crop yield based on long-term fertilization applications in Loess Plateau, China [J]. Plant Soil, 2015, 390 (1 – 2): 401 – 417.

植作物和饲养牲畜在内的多种农业生产活动。① 因此，仅仅关注单一农业生产部门与事实不相符，可能会削弱分析结果的完整性。

本章研究的贡献在于，分析黄土高原地区降水减少和农业碳税征收在小农户采取适应措施情况下的综合作用，以及小农户对这些综合作用的适应行为。本章利用一个全农场生物经济优化模型，探索了小农户可采用的多种适应措施的综合效果。使用全农场模型意味着本章研究能够分析农业系统中不同组成部分的系统变化，例如，模型能够将改变农地利用的适应措施包含在内。

7.3 研究对象与分析方法

7.3.1 研究区域

黄土高原地区（东经100.9~114.5度、北纬33.7~41.3度）位于中国北部，面积约为64万平方千米。该地区是中国也是东亚最重要的雨养农业区。当地人口约1.1亿，其中超过70%从事农业生产活动。② 黄土高原大约3/4的土地被用于农业生产。受西伯利亚高压与亚洲夏季季风活动的影响，当地的气候属于大陆性半干旱气候，夏季炎热湿润，冬季寒冷干燥。③ 由于存在严重的土壤侵蚀，当地环境十分脆弱。④ 虽然雨养农业也存在于青藏高原，然而受环境与社会经济因素的影响，青藏高原地区的农业生产规模要远

---

① Fu, B., Liu, Y., Lü, Y., et al. Assessing the soil erosion control service of ecosystems change in the Loess Plateau of China [J]. Ecological Complexity, 2011, 8 (4): 284-293. Tang, K., He, C., Ma, C., Wang, D. Does carbon farming provide a cost-effective option to mitigate GHG emissions? Evidence from China [J]. Australian Journal of Agricultural and Resource Economics, 2019, 63 (3): 575-592.

② 宁夏回族自治区统计局：《宁夏统计年鉴2015》，中国统计出版社2015年版。

③ Chen, H., Zhao, Y., Feng, H., Li, H., Sun, B. Assessment of climate change impacts on soil organic carbon and crop yield based on long-term fertilization applications in Loess Plateau, China [J]. Plant Soil, 2015, 390 (1-2): 401-417.

④ Fu, B., Wang, Y., Lu, Y., He, C., Chen, L., Song, C. The effects of land-use combinations on soil erosion: a case study in the Loess Plateau of China [J]. Progress in Physical Geography, 2009, 33 (6): 793-804. Feng, X., Fu, B., Piao, S., et al. Revegetation in China's Loess Plateau is approaching sustainable water resource limits [J]. Nature Climate Change, 2016, 6 (11): 1019-1022. Tang, K., He, C., Ma, C., Wang, D. Does carbon farming provide a cost-effective option to mitigate GHG emissions? Evidence from China [J]. Australian Journal of Agricultural and Resource Economics, 2019, 63 (3): 575-592.

小于黄土高原地区。

黄土高原地区的农业生产主要是以小农户经营的方式进行。当地农户农地面积多在 0.5~3 公顷，平均为 1.784 公顷。① 作物种植与牲畜饲养是当地最普遍的农业生产活动。类似的农业系统还存在于伊朗、哈萨克斯坦南部以及乌兹别克斯坦东部。超过 1/2 的农地被用于在 3~4 月间种植作物（如小麦）。剩余的农地用于种植牧草，为饲养的牲畜（如绵羊）提供饲料。该地区活羊供应量超过全国总量的 20%。② 高原浅层地下水贫乏，大部分地区地下水埋藏得很深，多在 60~70 米以下。③ 由于地表水和地下水的缺乏，当地农业生产主要依靠天然降水。

7.3.2　全农场土地利用优化模型

本章使用卡坦宁模型来分析气候变化对于雨养农业系统的影响。④ 卡坦宁模型是一个考虑了农地利用动态效果（如作物轮作对于未来产量的影响预期）的跨期农地利用优化模型。该模型的结构主要基于 MIDAS 模型的设计。MIDAS 模型是一个被广泛运用于雨养复合经营农业的静态全农场农地利用优化模型。⑤ 与之不同的是，卡坦宁模型是一个动态农地优化模型。该模型运用混合整数规划算法最大化农户净毛利，同时，基于农地管理单元利用历史，对农地利用序列选择进行优化。该模型考虑了雨养农业在环境、管理以及财

---

① 宁夏回族自治区统计局：《宁夏统计年鉴2015》，中国统计出版社2015年版。

② Tang, K., He, C., Ma, C., Wang, D. Does carbon farming provide a cost-effective option to mitigate GHG emissions? Evidence from China [J]. Australian Journal of Agricultural and Resource Economics, 2019, 63 (3): 575-592.

③ 田磊：《变化环境下黄土高原水文气候要素数值模拟及未来预测》，西北农林科技大学2019年博士学位论文。

④ Tang, K., Hailu, A., Kragt, M. E., Ma, C. The response of broadacre mixed crop-livestock farmers to agricultural greenhouse gas abatement incentives [J]. Agricultural Systems, 2018, 160: 11-20. Tang, K., He, C., Ma, C., Wang, D. Does carbon farming provide a cost-effective option to mitigate GHG emissions? Evidence from China [J]. Australian Journal of Agricultural and Resource Economics, 2019, 63 (3): 575-592.

⑤ Morrison, D. A., Kingwell, R. S., Pannell, D. J., Ewing, M. A. A mathematical programming model of a crop-livestock farm system [J]. Agricultural Systems, 1986, 20 (4): 243-268. Kragt, M. E., Pannell, D. J., Robertson, M. J., Thamo, T. Assessing costs of soil carbon sequestration by crop-livestock farmers in Western Australia [J]. Agricultural Systems, 2012, 112: 27-37. Thamo, T., Addai, D., Pannell, D. J., et al. Climate change impacts and farm-level adaptation: Economic analysis of a mixed cropping-livestock system [J]. Agricultural Systems, 2017, 150: 99-108.

务约束等方面的特征,涵盖了不同的作物、牧草以及畜牧(绵羊饲养)生产活动。绵羊的种群结构、能量需求以及能量供给(作物残茬、谷物以及牧草)由该模型直接模拟求出。模型结果反映了包括最优农地分配在内的农户最优经营决策组合。

最新版的卡坦宁模型包含了测算农业温室气体年排放量以及以碳税形式代表农业温室气体减排政策的模块。[①] 所使用的农业温室气体测算方法来自政府间气候变化专门委员会所公布的清单方法。[②] 测算方法的更多细节可参见唐凯等[③]的附录以及本书附录。所有的温室气体排放量依照唐凯等[④]的方法,使用全球暖化潜力值转化为 CO_2e。

本章模拟的是黄土高原地区的一个典型农户。农户耕地面积为 1.784 公顷。由于本地区农地通常包括多种土壤类型,本章研究考虑了四种具有不同生产特征的土壤(参见第 4 章)。农户从事种—畜复合经营获取利润。模型所涵盖的农地利用方式包括种植小麦、燕麦、油菜、豌豆以及豆科牧草(主要用于饲养绵羊)。考虑到当气候变化造成农业生产无法盈利时,农民可能选择暂时停止农业活动,本章将休耕纳入农地利用方式中。本章农业活动优化的周期为 10 年。

7.3.3 情景、价格以及模型有效性

许多最近发表的关于黄土高原地区的研究都认为,该地区气候存在变干的趋势。[⑤] 在陕西、宁夏、甘肃东部、青海东部、鄂尔多斯等区域减少趋势

---

①③ Tang, K., He, C., Ma, C., Wang, D. Does carbon farming provide a cost-effective option to mitigate GHG emissions? Evidence from China [J]. Australian Journal of Agricultural and Resource Economics, 2019, 63 (3): 575–592.

② 具体可参见 https://www.ipcc-nggip.iges.or.jp/public/2006gl/vol4.html.

④ Tang, K., Hailu, A., Kragt, M. E., Ma, C. Marginal abatement costs of greenhouse gas emissions: broadacre farming in the Great Southern Region of Western Australia [J]. Australian Journal of Agricultural and Resource Economics, 2016, 60 (3): 459–475.

⑤ Xin, Z., Yu, X., Li, Q., Lu, X. X. Spatiotemporal variation in rainfall erosivity on the Chinese Loess Plateau during the period 1956–2008 [J]. Regional Environmental Change, 2011, 11 (1): 149–159. Huang, J., Yu, H., Guan, X., Wang, G., Guo, R. Accelerated dryland expansion under climate change [J]. Nature Climate Change, 2016, 6 (2): 166–171. 程楠楠、何洪鸣、逯亚杰等:《黄土高原近 52 年降水时空动态特征》,载《山东农业大学学报(自然科学版)》2016 年第 3 期。

明显。① 20世纪后半叶以来，区域年平均降水量减少超过40毫米，降幅为7.9毫米/10年，年均降水量总体上下降了10%。② 1999~2008年平均降水量仅为392.85毫米，比多年平均值减少47.86毫米，下降10.86%。③ 黄土高原春小麦和春玉米各生育阶段干旱强度呈增加趋势；夏玉米干旱强度在陕西北部、宁夏和河西走廊呈增加趋势。④ 此外，预计降水量下降趋势还将持续。黄土高原地区年均降水量预计到2030年将下降5%~10%，到2050年将下降10%~30%⑤。此外，作为一种潜在的减缓气候变化影响的应对措施，有关学者也呼吁中国政府对包括黄土高原地区雨养复合经营农业在内的农业部门所排放的温室气体征收排放税。⑥

虽然黄土高原地区潜在气候变化的长期趋势基本清楚，目前对这些变化的潜在变化幅度还知之甚少。因此，本章将年度降水量以及农业碳税的变化情景设定在一个较大的范围（见表7-1）。类似的方法已被一些分析具有类似自然与环境条件的农业地区气候变化的研究所采用。⑦ 本章总共考虑30种情景，由6种年度降水减少幅度以及5种农业碳税税率所构成。这些情景一方面与已有研究的结论基本一致，另一方面也能反映气候变化与相关政策的不确定性，从而确保了分析结果的稳健性。

基准年均降水量设定为420毫米，以反映当地的平均降水水平。⑧ 本章

---

① 刘玉洁、陈巧敏、葛全胜等：《气候变化背景下1981~2010中国小麦物候变化时空分异》，载《中国科学：地球科学》2018年第7期。马雅丽、郭建平、赵俊芳：《晋北农牧交错带作物气候生产潜力分布特征及其对气候变化的响应》，载《生态学杂志》2019年第3期。

② 赵一飞、邹欣庆、张勃等：《黄土高原甘肃区降水变化与气候指数关系》，载《地理科学》2015年第10期。

③ 程杰：《黄土高原草地植被分布与气候响应特征》，西北农林科技大学2011年博士学位论文。

④ 何斌、刘志娟、杨晓光等：《气候变化背景下中国主要作物农业气象灾害时空分布特征（Ⅱ）：西北主要粮食作物干旱》，载《中国农业气象》2017年第1期。

⑤ 任婧宇、彭守璋、曹扬等：《1901~2014年黄土高原区域气候变化时空分布特征》，载《自然资源学报》2018年第4期。

⑥ Tang, K., Kragt, M. E., Hailu, A., Ma, C. Carbon farming economics: What have we learned? [J]. Journal of Environmental Management, 2016, 172: 49-57.

⑦ Thamo, T., Addai, D., Pannell, D. J., et al. Climate change impacts and farm-level adaptation: Economic analysis of a mixed cropping-livestock system [J]. Agricultural Systems, 2017, 150: 99-108. Tang, K., Hailu, A., Kragt, M. E., Ma, C. The response of broadacre mixed crop-livestock farmers to agricultural greenhouse gas abatement incentives [J]. Agricultural Systems, 2018, 160: 11-20.

⑧ Zhang, B., He, C., Burnham, M., Zhang, L. Evaluating the coupling effects of climate aridity and vegetation restoration on soil erosion over the Loess Plateau in China [J]. Science of the Total Environment, 2016, 539: 436-449.

在分析时考虑的是一种固定农业碳税。例如，情景 -10%/¥50t^{-1}CO$_2$e 表示年均降水量比基准降水量 420 毫米减少 10%[①]，且农业碳税为 50 元每吨 CO$_2$e。本章所使用的价格是 2015 年收购价。

表 7-1　　　　构成气候变化情景的年度降水量以及农业碳税变化

年度降水量减少幅度（%）	0	-5	-10	-15	-20	-30
农业碳税税率（元/tCO$_2$e）	0	50	100	150	200	—

卡坦宁模型自海努等[②]提出以来，已经被运用于分析一些国家的雨养农业系统。该模型被经常性地更新，以反映农业生态系统、资源、行为、技术、价格以及成本等方面的变化。[③] 为了验证所使用模型的有效性，一些对黄土高原地区农业系统有充分了解与经验的农业科学家和当地专家对模型参数、模拟结果以及模型的运行情况进行了校验。[④] 他们认为，模型的相关细节以及整体结果是合理的，符合所研究地区的实际情况。

7.4　研究结果

在基准情景中，最优化农业经营所产生的年均毛利为每公顷 3638 元。农业温室气体排放量为每公顷 2.35tCO$_2$e。约有 1/2 的农地用于种植作物，余下农地用于种植牧草。小麦是主要种植的作物。有 43.3% 的农地用于种植小麦，8.3% 的农地用于种植燕麦。小麦—牧草组合是最主要的作物轮作组合。

[①] 这些情景考虑了降水强度的变化。降水的其他特征保持不变。
[②] Hailu, A., Durkin, J., Sadler, R., Nordblom, T. L. Agent-based modelling study of shadow, saline water table management in the Katanning catchment, Western Australia [R]. Research Report for RIRDC Project No. PRJ-000578, 2011.
[③] Tang, K., Kragt, M. E., Hailu, A., Ma, C. Carbon farming economics: What have we learned? [J]. Journal of Environmental Management, 2016, 172: 49-57. Tang, K., Hailu, A., Kragt, M. E., Ma, C. The response of broadacre mixed crop-livestock farmers to agricultural greenhouse gas abatement incentives [J]. Agricultural Systems, 2018, 160: 11-20.
[④] 在此对华中农业大学张安录教授、董捷教授以及西北农林科技大学张蚌蚌博士的宝贵建议表示感谢。

7.4.1 气候变化情景下的农业利润

图 7-1 展示了不同情景中预期农户利润的结果。结果显示，即使农户采取了最优适应措施，减少的年均降水量以及农业碳税还是会造成农业利润的下降。在 30 种反映可能幅度变化的情景中，有 26 种情景的年均毛利损失相较于基准情景低于 10%。总体来看，年均降水量减少幅度越大，农业碳税税率越高，农民毛利的损失越大。

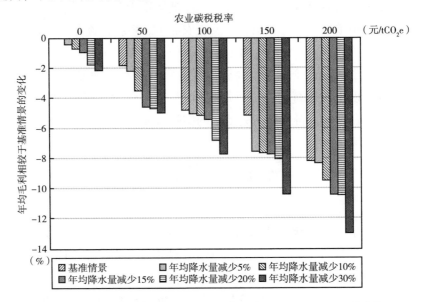

图 7-1　30 种情景中年均毛利相较于基准情景（6490 元）的变化

需要注意的是，在没有农业碳税的情况下，如果农民采取了最优适应措施，降水减少所造成的毛利损失最多 2%。当农业碳税税率增幅较大时，年均毛利会出现显著下降（见图 7-1）。

在没有出现降水量下降而需要缴纳农业碳税的情况下，200 元/tCO_2e 的农业碳税会造成年均毛利较基准情景下降 8%。当年均降水量减少 5% 时，农业利润的下降幅度较为轻微。在出现更极端的降水减少以及更高税率农业碳税的情况下，如果黄土高原地区的农民不采取有效的适应措施，农业利润会受到较大影响（如年均作物毛利会从每公顷 2600 元下降到 1200 元）。反之，如果农民采取了有效的适应措施，气候变化所造成的整体农业利润下降幅度较为温和，在大多数情况下会低于 10%。

7.4.2 农业温室气体排放量的变化

分析结果显示,在大多数情况下,农业温室气体排放量对于年均降水量以及农业碳税的变化是敏感的(见图7-2)。在出现降水减少的情况下,农民采取经济上最优的适应措施会增加温室气体的排放。如果降水减少的幅度是30%,农业温室气体排放量相较没有出现降水减少的情景会增加19%~49%;如果降水减少的幅度是5%,农业温室气体排放量的增加幅度在5%左右(见图7-3)。

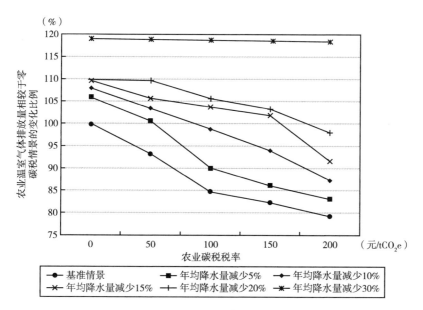

图7-2 30种情景中农业温室气体排放量相较于基准情景
(每公顷2.35tCO$_2$e)的变化

在出现不同幅度降水减少的情况下,黄土高原地区雨养农业农民对于农业碳税的响应大为不同。如果降水下降的幅度为5%或10%,农业碳税会导致农业温室气体排放的显著下降。在年均降水量下降幅度不高于10%的情况下,150元/tCO$_2$e的农业碳税会使农业温室气体排放量减少13%~17%,而200元/tCO$_2$e的农业碳税会使农业温室气体排放量减少约20%。然而,如果降水下降的幅度为30%,农业碳税不会造成农业温室气体排放的明显下降。即便农业碳税税率高达200元/tCO$_2$e,农业温室气体的下降幅度相较于零碳税情景也少于0.5%。

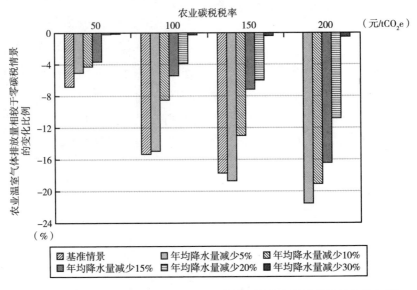

图7-3 30种情景中农业温室气体排放量相较于零碳税情景的变化比例

7.4.3 农民对于气候变化的适应

表7-2概括了6种情景下可以采取的优化后的农业经营决策或适应措施。选择这些情景的原因在于它们能够代表年均降水量下降以及农业碳税提高、从轻微到极端的变化影响。对气候变化的适应措施包括改变农地使用方式（农地分配、作物种类、作物轮作等）以及农业管理行为（畜群规模与结构等）。

经济上最优的农地利用通常对于潜在的变化较为敏感。当出现降雨减少的情况时，存在减少作物种植面积而增加牧草种植面积的趋势（见表7-2）。如果降水下降的幅度为5%~10%，增加同样幅度的农业碳税会增加作物种植面积。然而，如果降水下降的幅度为30%，农地分配对于农业碳税不敏感。此外，作物种植结构将由小麦主导型转向燕麦主导型。随着年度降水量的下降，优化后的农业经营决策将包括更多的燕麦—牧草轮作组合。休耕没有被优化后的农地利用所涵盖。

表 7-2 部分情景中优化后的农业经营决策组合

项目	单位	基准情景 0/0	所选择情景中相对于基准情景的变化（年度降水量减少（%）/农业碳税（元/tCO₂e））								
			0/50	0/200	5/0	5/50	5/150	15/150	20/150	30/150	30/200
年均毛利	元/公顷	3638	-29	-297	-15	-65	-273.62	-280	-292	-377	-471
作物种植面积	%	50.03	2.76	19.99	-0.4	-1.32	4.84	-2.42	-8.72	-10.65	-10.64
牧草种植面积	%	49.98	-2.77	-20	0.4	1.31	-4.83	2.41	8.71	10.64	10.63
小麦种植面积	%	43.31	1.55	15.45	-5.09	-16.66	-18.25	-21.84	-32.03	-41.85	-41.95
燕麦种植面积	%	8.25	1.02	2.07	4.65	13.81	19.14	17.89	20.35	29.51	29.78
氮肥施用量	kg	62.48	3.45	24.96	-0.5	-1.62	6.04	-2.96	-10.67	-13.12	-13.02
活畜年销售额	元	3080	330	-1360	447	776	401	1025	1399	2385	2405
羊毛年销售额	元	3090	330	-1390	1035	812	403	1025	1213	1799	1785
土壤类型1	轮作组合	PPW, PWW	WWP, PPP	OPO, PWPO	PPW, PWW	PWW, PPP	PPW, POW	PPW, POO	POO, PPP	OPO, PWPO	OPO, PWPO
土壤类型2	轮作组合	WPO, PO	WWP, POO, PWW	PPPP, POP, POO	WPO, PO	PWW, PPP, POO	PPP, POOP, PPP	PPP, POOP, PPP	PPP, POPW, OPW	PPPP, POP, POO	PPPP, POP, POO
土壤类型3	轮作组合	PWW, PWO, PPP	PPW, PWO, FWP	PPPP, POP, PPW	PWW, PWO, PPP	PWW, WPO	POO, POOP, PPP	POO, POOP, PPP	PPP, POP, OPPR	PPPP, POP	PPPP, POP
土壤类型4	轮作组合	WPO, POOP	WWP, PPP	OOPO, PPO	WPO, POOP	PWWR	PWO, POO	PWO, POO	PPP, POF	OOPO, PPO	OOPO, PPO

注：P—豆科牧草，W—小麦，O—燕麦，R—油菜，F—豌豆

除了农地利用模式，通过改变农业管理行为来适应气候变化也是必要的。经济上最优的绵羊种群规模①将会出现显著扩大。降水下降的幅度分别为10%、20%和30%时，种群规模将分别扩大26%、34%和68%。在作物种植措施方面，降水量减少时氮肥的施用也会减少。这表明，在降水量下降的情况下，种植部门会出现减产，相对应的肥料需求也会下降。

7.5　进一步讨论

考虑到气候变化潜在幅度与细节上的高度不确定性，气候变化对黄土高原地区农业利润的可能影响也不尽相同。但是，只要采取了气候变化适应措施，气候变化所造成的农业利润的损失便可以控制在一定的范围内。进一步地，气候变化负面影响所造成的农业收入上的损失，可能会由于农产品价格的上升而被抵消。对如中国这样人口众多的发展中国家而言，人口增长以及饮食结构转变的趋势极有可能会进一步扩大对农产品的需求。② 在农产品供给方面，剧烈的气候变化会威胁到农产品的有效生产。科学研究已发现，由这些因素所引起的农产品价格的上涨，可以在一定程度上缓解气候变化带来的负面影响。③

研究结果显示，在采取经济上最优的适应措施的情况下，降水量的下降会导致农业温室气体排放量的上升。在黄土高原地区，由于缺乏可用于灌溉的水源，降水量的显著减少会导致作物产量不可避免地下降，从而使种植业的利润也随之下降。而畜牧业的耐旱能力要远高于种植业，这意味着在降水量出现明显减少的情况下，畜牧业的盈利能力要高于种植业。因此，小农倾向于通过扩大畜牧业规模（如扩大畜群规模和增加牧草种植面积）来维持利润。由于畜牧业的温室气体排放密度更高，④ 畜牧业的扩张也会导致农业温

---

① 本书使用活畜年销售额与羊毛年销售额之和来简便估计畜群规模。

② Liu, Y., Fang, F., Li, Y. Key issues of land use in China and implications for policy making [J]. Land Use Policy, 2014, 40: 6–12. FAO. FAO Strategy on Climate Change [R]. FAO, Rome, 2017.

③ Wiebe, K., Lotze-Campen, H., Sands, R., et al. Climate change impacts on agriculture in 2050 under a range of plausible socioeconomic and emissions scenarios [J]. Environmental Research Letters, 2015, 10 (8), 085010.

④ Tang, K., Hailu, A., Kragt, M. E., Ma, C. Marginal abatement costs of greenhouse gas emissions: broadacre farming in the Great Southern Region of Western Australia [J]. Australian Journal of Agricultural and Resource Economics, 2016, 60 (3): 459–475.

室气体排放量的上升。

值得注意的是,小农户对于农业碳税的响应随着降水量下降幅度的不同而存在较大差异。当降水下降的幅度为5%~10%时,种植业受到的损失有限,利润水平下降的幅度小于10%。此外,由于种植业的排放密度要远小于畜牧业,因而种植业所需缴纳的农业碳税也远低于畜牧业,例如,在本章分析中,种植业每公顷温室气体排放量要比畜牧业低大约90%。因此,在降水下降的幅度为5%~10%的情景中,农业碳税的施行会引导小农户通过分配更多农地用于作物种植来扩大种植业生产,因而显著地减少了农业温室气体排放。

然而,若降水下降的幅度达到30%,作物产量会大幅下降,种植业的利润水平也随之显著降低。虽然畜牧业的外部成本更高,优化后的畜牧经营因其更强的耐旱能力而更具盈利能力方面的优势。[①] 即便农业碳税税率较高,黄土高原地区的小农户仍不会选择大幅缩小畜牧业的生产规模,以保证一定的利润水平。因此,在降水下降的幅度达到30%的情况下,农业碳税难以引起农业温室气体排放量的显著下降。

研究结果显示,当出现年均降水量下降时,小农户会选择采用更多的燕麦—牧草轮作组合以适应变化了的气候。如前所述,在气候变干的情况下,小农户会倾向于扩大畜牧业生产,因而牲畜的饲料需求会出现上升。除了种植更多的豆科牧草,农民也需要在冬末春初牧草无法满足饲料需求时,为畜群提供谷物饲料,以维持畜群的健康水平。在黄土高原地区,燕麦的亩产要比小麦高约40%。雨养复合经营农业系统在年均降水量减少的情况下,采用更多的燕麦—牧草轮作组合可以避免出现牲畜饲料供应短缺。因此,当地小农户将采用更多的燕麦—牧草轮作组合作为必要的气候变化适应措施。

在绝大多数情况下,休耕并没有成为经济上最优的农地利用方式。然而在现实中,考虑到其他环境政策的实施,情况可能会有所不同。例如,我国中央政府和许多地方政府已经推出了相应的政策鼓励休耕,其中也包括了黄土高原地区的一些地方政府。[②] 这些旨在减少水资源消耗以及土壤污染的政策为采取休耕措施的农民提供了额外的收入。因此,如果综合考虑这些政策,

---

① Thamo, T., Addai, D., Pannell, D. J., et al. Climate change impacts and farm-level adaptation: Economic analysis of a mixed cropping-livestock system [J]. Agricultural Systems, 2017, 150: 99–108.

② 参见 https://www.agcanada.com/daily/china-to-let-land-lie-fallow-as-grain-stocks-surge.

休耕也能够成为经济上最优农地利用的一种方式,这样也有助于传统农业朝着可持续化转变。

研究结果显示,如果降水下降的幅度为5%或10%,农业碳税会导致农业温室气体排放的显著下降。在年均降水量下降幅度不高于10%的情况下,150元/tCO_2e的农业碳税会使得农业温室气体排放量减少13%~17%,而200元/tCO_2e的农业碳税会使得农业温室气体排放量减少约20%。当温室气体边际减排成本低于农业碳税时,农民会意识到排放与减排相比,前者更为昂贵,此时他们便会选择减少温室气体排放。[1] 因此,以上结果意味着在黄土高原地区减少13%~17%农业温室气体排放的边际减排成本低于150元/tCO_2e,而减少20%农业温室气体排放的边际减排成本低于200元/tCO_2e。

考虑到实施的碳交易计划以及农业温室气体排放在全国排放总量当中所占的比重,[2] 测算与理解中国农业温室气体边际减排成本是十分重要的。这将有助于确定将哪些具有经济竞争性的行业纳入碳交易计划中。近年来,有学者分析了中国的制造业、火电行业、城市工业以及钢铁行业。他们发现,这些行业碳排放的边际减排成本在298元/tCO_2e至17500元/tCO_2e之间,明显高于本章的结果。[3] 这意味着,若被纳入全国性的碳交易机制中,黄土高原地区的雨养复合经营农业能够减少更多的温室气体排放。这样,黄土高原地区的小农户能够通过出售碳信用的方式,将其出售给那些边际减排成本更高的市场主体,以获得额外的收入,提高农业适应活动的盈利能力。

与其他研究相比,本章所采用的全农场生物经济分析方法能够让模拟结果更为精确。现有关于黄土高原地区气候变化作用的研究多关注单一农业生产部门,忽略了当地农业系统中众多组成部分的综合作用。然而,这些综合作用往往会对雨养复合经营农业产生影响。例如,种植更多的豆科牧草能够为后茬谷类作物的生长提供更多的氮,而调整作物轮作组合能够改变用作牲畜饲料的残茬的数量。改进版的卡坦宁模型考虑了这些农业系统中的复杂关系。此外,绝大多数的相关文献只考虑了小麦。而在极端气候条件下,小麦

---

[1] Tang, K., Hailu, A., Kragt, M.E., Ma, C. The response of broadacre mixed crop-livestock farmers to agricultural greenhouse gas abatement incentives [J]. Agricultural Systems, 2018, 160: 11-20.

[2] Tang, K., He, C., Ma, C., Wang, D. Does carbon farming provide a cost-effective option to mitigate GHG emissions? Evidence from China [J]. Australian Journal of Agricultural and Resource Economics, 2019, 63 (3): 575-592.

[3] Wang, K., Che, L., Ma, C., Wei, Y. The shadow price of CO_2 emissions in China's iron and steel industry [J]. Science of the Total Environment, 2017, 598: 272-281.

的适应能力要弱于许多其他作物。① 进一步的，绝大多数的相关文献使用的是生物物理学的方法，忽略了生产利润变化对于农民经营决策的影响。

本章的研究还存在一些不足。例如，大气中 CO_2 浓度的增加可能会提高作物的产量，② 而本章所使用的模型并没有考虑这一点。此外，分析中所考虑的作物与牧草都是黄土高原地区现有的典型种植品种，这些品种的生长对于降水的变化异常敏感。新品种的引入以及农业生产管理的改善也许能够增强作物与牧草的耐旱能力。未来的研究需要在农民对气候变化以及乡村可持续性的适应活动方面，进行更全面更深入的分析。

7.6 本章小结

本章分析了在考虑适应措施的情况下，气候变化对中国黄土高原地区小农户在农业利润、农业生产以及相关温室气体排放方面的影响。本章使用了一个全农场生物经济优化模型。为了反映气候变化以及相关政策的不确定性，本章考虑了一系列的年度降水量以及农业碳税变化情景，从而确保了分析结果的稳健性。

研究结果显示，即使农户采取了最优适应措施，减少的年均降水量以及农业碳税还是会造成农业利润的下降。然而，如果采取了最优适应措施，气候变化所造成的小农户利润损失可以控制在一定的范围之内。

如果出现降水减少的情况，农民采取经济上最优的适应措施会改变农地使用方式，增加温室气体的排放。若降水减少了30%，与没有出现年度降水量下降的情况相比，农业温室气体排放量会增加19%~49%。如果降水减少的幅度是5%，农业温室气体排放量的增加幅度在5%左右。在降水下降的幅度为5%或10%，在此情况下，农业碳税会导致农业温室气体排放的显著下降。然而，如果降水下降的幅度为30%，农业碳税不太可能造成农业温室气

---

① Simelton, E., Fraser, E. D., Termansen, M., et al. The socioeconomics of food crop production and climate change vulnerability: a global scale quantitative analysis of how grain crops are sensitive to drought [J]. Food Security, 2012, 4 (2): 163-179. Albers, H., Gornott, C., Hüttel, S. How do inputs and weather drive wheat yield volatility? The example of Germany [J]. Food Policy, 2017, 70: 50-61.

② Thamo, T., Addai, D., Pannell, D. J., et al. Climate change impacts and farm-level adaptation: Economic analysis of a mixed cropping-livestock system [J]. Agricultural Systems, 2017, 150: 99-108.

体排放的明显下降。

经济上最优的农地利用通常对于潜在的变化较为敏感。当出现降雨减少的情况时，存在减少作物种植面积而增加牧草种植面积的趋势。如果降水下降的幅度为5%~10%，增加同样幅度的农业碳税会增加作物种植面积。然而，如果降水下降的幅度为30%，农地分配对于农业碳税不敏感。此外，作物种植结构将由小麦主导型转向燕麦主导型。随着年度降水量的下降，优化后的农业经营决策将包括更多的燕麦—牧草轮作组合。休耕没有被优化后的农地利用所涵盖。除了农地利用模式，通过改变农业管理行为来适应气候变化也是必要的。经济上最优的绵羊种群规模将会出现显著扩大。

本章的分析结果具有重要的政策启示。首先，政策制定者需要为小农户提供更多的农业推广培训项目，引导小农户充分合理地利用现有的气候变化适应措施。这样可以显著降低气候变化所造成的农业利润损失。其次，在制定与实施农业政策时，需要充分考虑气候方面的变化，如降水量等。这样能够提高农业政策的有效性，促进农村农地利用改革，优化全社会的福利。最后，可以考虑将小农户纳入全国性的碳交易机制中。小农户能够通过出售碳信用的方式，将其出售给那些边际减排成本更高的市场主体，以获得额外的收入，提高农业适应活动的盈利能力，缩小城乡收入差距。

第8章 结论、建议与展望

8.1 全书总结

气候变化对全球自然生态和人类系统构成严峻挑战。为了应对气候变化给全球带来的挑战,国际社会开始采取联合行动来抑制温室气体的排放,积极主动地进行适应。雨养农业是世界上最大、受气候变化影响最深远最直接的农业生产部门。气候变化增加了全球雨养农业区农业生产的不确定性,直接影响了农业生产的布局和结构。因此,有必要弄清气候变化所带来的自然环境以及社会经济等方面变化对雨养农业生产的影响,并在此基础上探寻雨养农业对气候变化的适应性路径。这是合理预测未来可持续发展必不可少的工作,对于包括中国在内的世界各国特别是广大发展中国家应对气候变化、保障农业生产与粮食安全、减少农村贫困、促进区域环境经济协同发展有着重要意义,也是当前学界较少涉及的一个新领域。本书正是基于这样的背景而开展的。

本书从雨养农业的具体特点和实际出发,基于已有的科学研究成果和实地观测数据,系统梳理雨养农业区气候变化的中长期趋势及其潜在变化幅度,分析气候变化对雨养农业生产的影响,通过国际比较,从微观的视角出发,分析雨养农业温室气体减排的成本有效性,探索在优化农户收益前提下雨养农业对气候变化的有效适应路径,为应对气候变化、保障农业生产与粮食安全、推动农业绿色发展与转型、促进区域环境经济协同发展提供理论依据和政策参考。

研究结论如下。

首先,在雨养农业部门中减少温室气体排放是一项相对有效且低成本的选择。在澳大利亚大南区，2006~2013年雨养农业年均温室气体减排潜力值在14%~33%波动，均值为21%。样本农场平均影子价格为17.6澳元/tCO_2e（约合88元/tCO_2e）。实现减少13%温室气体减排目标的边际减排成本不高于20澳元/tCO_2e（约合100元/tCO_2e），而减少36%排放的边际减排成本低于30澳元/tCO_2e（约合150元/tCO_2e）。在中国黄土高原地区，种—畜复合经营小农减少16.6%和33%温室气体排放的边际减排成本，以2015年人民币计算不高于100元/tCO_2e和150元/tCO_2e。总体来看，在雨养农业部门进行温室气体减排活动具有成本有效性上的一定优势。

其次，在澳大利亚大南区和中国黄土高原地区，即使农户采取了最优适应措施，气候变化的影响还是会造成农业利润的下降。如果采取了最优适应措施，气候变化所造成的利润损失可以控制在一定的范围之内。在澳大利亚大南区，农民在农业碳税税率为20澳元/tCO_2e的情况下，优化后的种—畜生产组合的每公顷毛利相较零碳税时减少4%。进一步地，若农业碳税税率为30澳元/tCO_2e，与零碳税情景相比每公顷毛利润减少的幅度低于8%。在黄土高原地区，在30种反映可能幅度变化的情景中，有26种情景的年均毛利损失相较于基准情景低于10%。在没有农业碳税的情况下，如果农民采取了最优适应措施，降水减少所造成的毛利损失最多2%。在没有出现降水量下降而需缴纳农业碳税的情况下，200元/tCO_2e的农业碳税会造成年均毛利较基准情景下降8%。当年均降水量减少5%时，农业利润的下降幅度较为轻微。在出现更极端的降水减少以及更高税率农业碳税的情况下，如果农民采取了有效的适应措施，气候变化所造成的整体农业利润下降幅度较为温和，在大多数情况下会低于10%。

再次，经济上最优的农地利用通常对于潜在的变化较为敏感。由于牲畜是主要的农业温室气体排放源，在优化后的雨养复合农业系统中，种植业的比例将上升，农业系统的结构会朝着作物主导型发展。受农业碳税的影响，在优化后的生产经营组合中，农民倾向于包含更多更加适应气候变化作物的轮作组合。如果出现降水减少的情况，农民所采取经济上最优的适应措施会改变农地使用方式，增加温室气体的排放。在澳大利亚大南区，当农业碳税税率不断提高时，农民倾向于在种植作物时减少包含油菜的轮作组合的使用频率，而更多地选择包含紫花豌豆的轮作组合。在中国黄土高原地区，当出现降雨减少的情况时，存在减少作物种植面积而增加牧草种植面积的趋势。如果降水下降的幅度为5%~10%，增加同样幅度的农业碳税会增加作物种植面积。然而，如果降水下降的幅度为30%，农地分配对于农业碳税不敏感。此外，作物种植结构将由小麦主导型转向燕麦主导型。随着年度降水量的下降，优化后的农业经营决策将包括更多的燕麦—牧草轮作组合。休耕没有被优化后的农地利用所涵盖。经济上最优的绵羊种群规模将会出现显著扩大。降水下降的幅度分别为10%、20%和30%时，种群规模将分别扩大26%、34%和68%。在作物种植措施方面，降水量减少时氮肥的施用也会减少。

最后，随着农业碳税税率的增加，农业利润以及农业温室气体排放量均出现下降。然而，相对较低的碳税税率，可以实现农业温室气体的有效减排，且减排成本相对较小。在澳大利亚大南区，20澳元/tCO_2e（约合100元/tCO_2e）的碳税可以减少13%农业温室气体排放，30澳元/tCO_2e（约合150元/tCO_2e）的碳税可以减少30%农业温室气体排放。在中国黄土高原地区，150元/tCO_2e的碳税可以减少13%~17%农业温室气体排放，200元/tCO_2e的碳税可以减少20%农业温室气体排放。

8.2 政策建议

本书的分析结果具有重要的政策启示，具体如下。

雨养农业能够在显著减少温室气体排放量的同时，节约农业生产所需的投入。经营雨养农业的农民能够通过提高生产效率，将其现有生产朝着经济—环境双赢的局面转变，同时实现生产增收以及对气候变化的有效应对。相关政府部门应当考虑改进现有以及设计更多的政策工具，以促进雨养农业减少其生产中的技术无效水平，例如，修改那些限制雨养农业经营灵活性的不适宜的政策规定。政府也可以考虑为提升雨养农业生产效率，以及提升家庭农场经营管理水平，制订和实施相应的促进政策和计划。

考虑到中国以及澳大利亚近期试行的市场化碳减排的经验，在雨养种—畜复合农业中进行碳汇农业，是一项极具成本有效性优势的选择。为半干旱雨养种—畜复合农业部门提供农业温室气体减排政策，不仅能够通过改变农户农地利用和农业生产行为，显著减少农业温室气体排放，还可以以极具成本有效性优势的方式，实现全社会的有效减排，推动减碳目标的早日实现。

若雨养农业被纳入碳税系统中，则所测算的农业温室气体影子价格或边际减排成本可以作为碳税税率的一个参考值。政府在筹建排放交易市场时，可以以所测算的影子价格或边际减排成本为基准，设定排放额度的初始价格；可以考虑以温室气体排放量作为权重的加权平均的温室气体影子价格作为排放额度初始价格的参考，结合不同地区排放成本的差异，制定区域减排单位的折算制度。

政府可以考虑将小农户纳入全国性的碳交易机制中。小农户能够通过出售碳信用的方式，将其出售给那些边际减排成本更高的市场主体，以获得额外的收入，提高农业适应活动的盈利能力，缩小城乡收入差距。

政府需要为雨养农业设计动态的减排定价政策。如果减排的机会成本无法得到完全补偿，雨养农业农民将不愿进行农业温室气体减排活动。因此，政府可以考虑动态修订碳税税率，以及合理地设定温室气体交易市场的排放权初始价格，以反映减排机会成本的波动。

政策制定者需要为小农户提供更多的农业推广培训项目，引导小农户充分合理地利用现有的气候变化适应措施。这样可以显著降低气候变化所造成的农业利润损失。

在制定与实施农业政策时，需要充分考虑气候以及社会方面的变化，如降水量和农户经营特征等。这样能够提高农业政策的有效性、促进农村农地利用改革、优化全社会的福利。

气候变化主管部门要从战略全局认识和把握应对气候变化目标任务，坚定不移持续实施积极应对气候变化的国家战略。要充分发挥生态环境部门牵头协调作用，会同有关部门采取强有力措施，全面推进结构调整和绿色发展，优化经济结构和产业结构，继续控制化石能源消费，大力发展非化石能源，深化能源和价格改革，在农业、林业、土地利用、草原、湿地等方面实施"基于自然的解决方案"，加强生态环境的保护、治理和修复，提升生态系统的服务功能，增加碳汇，推动形成应对气候变化工作的强大合力，切实提升气候治理能力，以更加积极地应对气候变化行动，为应对全球气候变化作出积极贡献。

中国生态环境部门作为气候变化主管部门，可以从以下方面深入推进应对气候变化工作。

一是积极推动落实"十四五"应对气候变化目标任务。研究实现习近平总书记重大对外宣示相衔接的"十四五"碳强度指标与其他应对气候目标任务的具体路径；编制实施"十四五"应对气候变化专项规划，并建立有效的监督考核机制；切实提高各级党委、政府落实应对气候变化目标任务的主动性和自觉性。

二是积极开展二氧化碳排放达峰行动。围绕落实习近平总书记宣布的新达峰目标，制订二氧化碳排放达峰行动计划，在"十四五""十五五"期间，推动地方、部门和重点行业开展达峰行动，确保如期实现达峰目标。

三是加快推进低碳发展重点工作。努力克服疫情等不利因素影响，加快推进全国碳市场制度建设、系统建设和基础能力建设，争取尽快实现上线交易。继续推进低碳试点示范，支持有条件的地方开展近零碳甚至零碳示范区建设，加快启动气候投融资试点。

四是加强适应气候变化工作。组织编制《国家适应气候战略2035》，部署相关领域适应气候变化工作，提升重点区域适应能力，深化气候适应型城市试点。

五是积极参与全球气候治理。继续坚持多边主义，坚持共同但有区别的责任原则，持续推进气候变化国际谈判，推动《巴黎协定》的全面、有效、持续实施。根据《巴黎协定》的有关要求，组织编制落实国家自主贡献进展报告和21世纪中叶长期温室气体低排放发展战略并按时提交。加强与各国应对气候变化领域的交流合作，重点推进气候变化南南合作。

六是提升全社会应对气候变化意识。利用好"全国低碳日""中国角"系列边会等宣传活动，拓宽宣传渠道，普及应对气候变化知识，宣传应对气候变化理念，讲好应对气候变化的"中国故事"。

8.3 研究展望

在解读本书所得出的结论时需注意以下问题。

第一，本书对农业温室气体排放量的估计，是建立在有关农产品价格、土壤类型以及气候条件的一系列假设基础之上的。本书的结果，是基于处于半干旱气候条件下、以种—畜复合农业为代表的雨养农业。研究结果也许能够用来理解在中亚东南部地区、伊朗高原西部以及南部非洲一些内陆地区类似农业系统中的农业减排。然而，中国及一些国家其他的农业区域在农业系统、土壤类型和气候上的差异，以及农产品市场价格的波动，都有可能对农民在农地利用、生产毛利以及温室气体排放方面的响应产生影响，需要未来进一步研究。

第二，本书没有考虑由农业减排活动所带来的潜在的协同效益，如水土流失的减缓以及土壤品质的改良等。① 许多雨养农业地区（如黄土高原地区）存在着严重的土壤侵蚀和地表径流，土壤通常肥力较低。农民可以通过采取如轮作和保护性耕作在内的气候变化应对措施，减少土壤侵蚀、改进土壤肥力。类似私人的协同效应能够进一步地降低温室气体排放的边际减排成本。此外，一些气候变化应对措施也能够带来其他的公共效用，如生态多样性保护以及一些社区福利的改善等。对于这些私人和公共协同效应的综合分析说明，政府应当提供更优厚的农业温室气体减排政策。未来的研究应当考虑这些潜在的协同效应。

第三，值得注意的是，一些实证研究认为，雨养农业部门采取气候变化应对措施会产生一系列与农业生产非直接相关的额外成本，如交易成本和学习成本等。② 本书所使用的模型没有考虑这些额外成本。然而在实际操作中，这些额外成本可能给经营种—畜复合农业农民采取气候变化应对措施带来一定的阻碍。未来的研究有必要全面考虑气候变化应对过程中的成本因素。

第四，本书所探讨的农业温室气体减排政策是一种简单的农业碳排放税，即农民需要为其在生产中所排放的每吨温室气体缴纳一定金额的税，无免税排放额，且税率固定。对于在不同类型农业温室气体减排政策背景下（如累进制排放税）农民的生产经营行为还需进一步探讨。

第五，除了气候变动所带来的风险，还需要进一步考虑其他风险对于模拟结果所产生的影响。未来的研究需要进一步地分析和模拟农民风险偏好以及农产品市场价格变动情况对农业温室气体减排政策响应所带来的影响。

第六，本书所使用的全农场生物经济学模型还存在一些不足。例如，大气中二氧化碳浓度的增加可能会提高作物的产量，③ 而本书所使用的模型并

---

① Tang, K., Kragt, M. E., Hailu, A., Ma, C. Carbon farming economics: What have we learned? [J]. Journal of Environmental Management, 2016, 172: 49 – 57.

② Bakam, I., Balana, B. B., Matthews, R. Cost-effectiveness analysis of policy instruments for greenhouse gas emission mitigation in the agricultural sector [J]. Journal of Environmental Management, 2012, 112: 33 – 44. Tang, K., Kragt, M. E., Hailu, A., Ma, C. Carbon farming economics: What have we learned? [J]. Journal of Environmental Management, 2016, 172: 49 – 57.

③ Thamo, T., Addai, D., Pannell, D. J., et al. Climate change impacts and farm-level adaptation: Economic analysis of a mixed cropping-livestock system [J]. Agricultural Systems, 2017, 150: 99 – 108.

没有考虑这一点。此外，分析中所考虑的作物与牧草都是黄土高原地区现有的典型种植品种，这些品种的生长对于降水的变化异常敏感。新品种的引入以及农业生产管理的改善也许能够增强作物与牧草的耐旱能力。未来的研究需要在农民对气候变化以及乡村可持续性的适应活动方面，进行更全面更深入的分析。

第七，本书没有具体考虑碳中和目标。碳中和是指市场主体测算在一定时间内直接或间接产生的温室气体排放总量，通过植树造林、节能减排等形式，抵消自身产生的二氧化碳排放量，实现二氧化碳"零排放"。要达到碳中和，一般有两种方法：一是通过特殊的方式去除温室气体，如生物固碳；二是减少碳排放。中国最新的气候行动目标，勾勒出中国低碳转型发展未来40年的远景宏图，也为"十四五""十五五"规划期间的低碳绿色领域发展提供指引。未来的研究可以根据碳中和目标时间表规划未来的技术路线图，谋划加强雨养农业转型升级，加快结构性低碳转变的具体路径，进一步明确雨养农业生物固碳以及碳减排的具体方式。

附　录

附录一　农业温室气体排放量计算方法*

（一）作物排放

雨养农业区气候往往具有低湿度、高温与强太阳辐射的特征，由作物秸秆产生的 N_2O 排放条件受到抑制，产生的 N_2O 应低于 IPCC 推荐值。参考唐凯的研究，本书对于由作物秸秆产生 N_2O 排放估计使用的排放因子值为 0.001。[①]

由秸秆还田所产生的 N_2O（R）：

$$R(\text{Gg } N_2O) = P \times R \times DM \times CC \times NC \times (1 - F - FFOD) \times EF \times C_g$$

其中：P 表示年度作物产量（Gg）；R 表示作物秸秆率（kg 作物秸秆/kg 作物）；DM 表示干物质含量（kg 干重/kg 作物秸秆）；CC 表示作物秸秆中碳的质量比例；NC 表示作物秸秆中氮对碳的比例；F 表示作物被燃烧的比例；$FFOD$ 表示作物被移走的比例；$EF = 0.001$；$C_g = 44/28$。

（二）化肥施用排放

产生自氮肥施用的氮排放（$F1$）：

---

* 本部分主要内容参考唐凯：《基于生物经济学的澳大利亚农业温室气体减排潜能分析》，人民出版社 2018 年版。

① Barton, L., Butterbach-Bahl, K., Kiese, R. et al. Nitrous oxide fluxes from a grain-legume crop (narrow-leafed lupin) grown in a semiarid climate [J]. Global Change Biology, 2011, 17: 1153 – 1166.

$$F1(GgN_2O) = 施用到生产系统中的化肥质量(GgN) \times EF \times C_g$$

其中：$EF = 0.001$；$C_g = 44/28$。

产生自大气沉降的氮排放（$F2$）：

$$F2(GgN_2O) = 化肥总质量(GgN) \times FracGASF \times EF \times C_g$$

其中：$FracGASF = 0.1$（IPCC 推荐值）；$EF = 0.01$；$C_g = 44/28$。

产生自淋溶和流失的氮排放（$F3$）：

$$F3(GgN_2O) = 生产系统中化肥的总质量(GgN) \times FracWET \\ \times FracLEACH \times EF \times C_g$$

其中：$FracWET$ 表示可供淋溶和流失的氮的比例；$FracLEACH = 0.3$（IPCC 推荐值）；$EF = 0.0125$；$C_g = 44/28$。

（三）畜牧排放

绵羊是澳大利亚与中国雨养农业中的主要牲畜。只有少部分的农场牧养了肉用牛。羊群可以被划分为以下几类：公羊（rams）、阉羊（wethers）、幼年母羊（maiden ewes）、成年母羊①（breeding ewes）、其他类母羊、羊羔和年龄在 1~2 岁之间的幼羊（hoggets）。牛群可分为公牛、母牛和阉牛。先计算不同种类的牲畜每头所产生温室气体排放的数量，然后乘以各种类的头数再进行加总，最终得到农场规模的排放量。畜牧排放所产生的 CH_4 主要来自羊与牛的肠道发酵和粪便。所产生的 N_2O 包括来自排泄到土壤中尿液与粪便的直接排放，由土壤中氮的淋溶和流失以及大气沉降所产生的间接排放。

产生自肠道发酵的 CH_4 排放（M）。

羊：

$$M(kg\ CH_4/只/天) = I \times 0.0188 + 0.00158$$

其中：I 表示每一类羊每天消耗的干物质（dry matter）（kg/只）。

肉用牛：

$$M(kg\ CH_4/天) = Y/100 \times GEI/F$$

---

① 成年母羊是指具有繁殖能力的母羊，幼年母羊尚不能繁殖。

其中：GEI 表示每日摄入的总能量（假设每 kg 干物质所含能量为 18.4MJ）；Y 表示摄入的总能量（GEI）转化为 CH_4 的百分比；$F = 55.22\text{MJ/kg } CH_4$。

产生自牲畜粪便的 CH_4 排放（M）。

羊与牛：

$$M(\text{kg } CH_4/只/天) = DMM \times MEF$$

其中：$DMM = I \times (1 - DMD)$；I 表示总摄入量（kg 干物质/只/天）；DMD 表示饲料的消化吸收率（%）；MEF 表示 CH_4 排放因子（1.4×10^{-5} kg CH_4 干物质粪便）；

（四）N_2O 排放

产生自土壤直接排放的 N_2O。

对于土壤中动物排泄物所产生的 N_2O 排放量的估计，国家温室气体清单（NGGI）2010 使用了由国际上相关研究所测得的排放因子。然而这些排放因子没有很好地反映样雨养农业区所具有的半干旱农业气候条件特征。[1] 目前缺乏相关地区 N_2O 排放的直接实验数据。

在雨养农业区，由于低湿度、高温与强太阳辐射，粪便变成厌氧状态的可能性降低。厌氧条件的不充分会限制 N_2O 的产生，[2] 这意味着该地区的排放因子应低于 NGGI 2010 所提供的推荐值。扎莫等提出在西澳的中部谷物带（Central Wheatbelt）地区牲畜尿液的排放因子为 0.14%，粪便为 0.01%。[3] 考虑到样本地区有着类似的气候、土壤以及农业生产特征，本书推断该地区动物排泄物所产生的 N_2O 的排放因子应当接近于中部谷物带的排放因子。因而，本书借用西澳中部谷物带的排放因子来计算土壤直接排放的 N_2O。

产生自尿液的 $N_2O(\text{Gg})$ = 排放到土壤中的尿液所含 $N \times 0.14\% \times C_g$

---

[1] Barton, L., Butterbach-Bahl, K., Kiese, R. et al. Nitrous oxide fluxes from a grain-legume crop (narrow-leafed lupin) grown in a semiarid climate [J]. Global Change Biology, 2011, 17: 1153 – 1166.

[2] van der Weerden, T. J., Luo, J., de Klein, C. A. M, et al. Disaggregating nitrous oxide emission factors for ruminant urine and dung deposited onto pastoral soils [J]. Agriculture, Ecosystems & Environment, 2011, 141: 426 – 436.

[3] Thamo, T., Kingwell, R. S., Pannell, D. J. Measurement of greenhouse gas emissions from agriculture: Economic implications for policy and agricultural producers [J]. Australian Journal of Agricultural and Resource Economics, 2013, 57: 234 – 252.

产生自粪便的 N_2O(Gg) = 排放到土壤中的粪便所含 $N \times 0.01\% \times C_g$

其中：$C_g = 44/28$。

产生自土壤间接排放的 N_2O（大气沉降）(D)。

$$D(GgN_2O) = 农业管理系统中挥发的氮(N) \times 0.01\% \times C_g$$

其中：$C_g = 44/28$。

淋溶和流失所产生的 N_2O (E)。

莉等（Li et al.）认为，排放因子值 0.001 适合西澳坎德林（Cunderdin）地区 37 年的气象数据所显示的气候条件。[①] 由于雨养农业地区多有着与坎德林相近的气候、土壤和农业生产特征，本书将该排放因子设置为 0.001。

$$E = M \times EF \times C_g$$

其中：E 表示淋溶和流失所产生的 N_2O 年排放量；M 表示通过淋溶和流失所流失的 N（Gg N）；$EF = 0.001$；$C_g = 44/28$。

（五）固氮作物排放

目前缺乏在雨养农业系统中固氮作物所产生的 N_2O 排放数据。因此，我们使用 NGGI 2010 给出的澳大利亚非灌溉牧草生长系统中施用化肥所产生的 N_2O 的排放因子（0.004）来替换推荐值。

固氮作物所产生的 N_2O (C)：

$$C(Gg\ N_2O) = P \times R \times DM \times CC \times NC \times EF \times C_g$$

其中：P 表示年度作物产量（Gg）；R 表示作物秸秆率（kg 作物秸秆/kg 作物）；DM 表示干物质含量（kg 干重/kg 作物秸秆）；CC 表示作物秸秆中碳的质量比例；NC 表示作物秸秆中氮对碳的比例；$EF = 0.004$；$C_g = 44/28$。

综上，这里主要对 NGGI 2010 提供的部分 N_2O 排放因子进行了替换，相关总结见附表 1。

[①] Li, Y., Barton, L., Chen, D. Simulating response of N_2O emissions to fertiliser N application and climatic variability from a rain-fed and wheat-cropped soil in Western Australia [J]. Journal of the Science of Food and Agriculture, 2011, 92: 1130 – 1143.

附表1　　　　　　　　　　N₂O 排放因子修正说明

排放源	替换原因	排放系数	来源
作物秸秆	雨养农业区气候具有低湿度、高温与强太阳辐射的特征，由作物秸秆产生的 N_2O 排放条件受到抑制，产生的 N_2O 应低于推荐值。	0.1%	Barton et al. (2011)[①]
土壤中牲畜排泄物	雨养农业区气候具有低湿度、高温与强太阳辐射的特征。该环境中排泄物变成厌氧状态的可能性降低，限制 N_2O 的产生，意味着该地区的排放因子应低于 NGGI 所提供的推荐值。当地气候特征也使得土壤中排泄物淋溶和流失的条件不够充分，故所产生的 N_2O 应低于推荐值。	0.14%（尿液） 0.01%（粪便） 0.1%（淋溶和流失）	Thamo et al. (2013)[②] Li et al. (2011)[③]
固氮作物	目前缺乏雨养农业区固氮作物所产生的 N_2O 排放数据，故使用 NGGI 澳大利亚非灌溉牧草生长系统中施用化肥所产生的 N_2O 的排放因子来替换。	0.4%	Natioanl Inventory Report 2010[④]

① Barton, L., Butterbach-Bahl, K., Kiese, R. et al. Nitrous oxide fluxes from a grain-legume crop (narrow-leafed lupin) grown in a semiarid climate [J]. Global Change Biology, 2011, 17: 1153-1166.

② Thamo, T., Kingwell, R. S., Pannell, D. J. Measurement of greenhouse gas emissions from agriculture: economic implications for policy and agricultural producers [J]. Australian Journal of Agricultural and Resource Economics, 2013, 57: 234-252.

③ Li, Y., Barton, L., Chen, D. Simulating response of N_2O emissions to fertiliser N application and climatic variability from a rain-fed and wheat-cropped soil in Western Australia [J]. Journal of the Science of Food and Agriculture, 2011, 92: 1130-1143.

④ Department of Climate Change and Energy Efficiency. National Inventory Report 2010 Vol. 1 [R]. Commonwealth of Australia, 2012.

附录二　全农场模型概述[*]

所谓全农场模型（whole-farm model，WFM），是指将农场生产作为一个系统，基于设定的优化目标，综合考量系统的生物、物理、环境、管理、财务以及科技方面的条件，寻求最优解的生物经济学模型。全农场模型能够帮

---

[*] 本部分主要内容引自唐凯：《基于生物经济学的澳大利亚农业温室气体减排潜能分析》，人民出版社 2018 年版。

助了解发生在农业系统内变化所造成的经济后果。① 依据罗伯森等（Robertson et al.）所提出的分类，目前学界使用的全农场模型大致可分为工业化农业系统中的静态优化模型（static optimisation in industrialised agriculture）、发展中国家家庭农业模型（household models in developing world agriculture）、生物物理学仿真模型（biophysical simulation）、整合静态优化与动态模拟的模型（integrate static optimisation with dynamic simulation）和结合农场实地调查的全农场模型（involves the use of WFM's with farm surveys）。②

工业化农业系统中的静态优化模型利用农场在生物、物理、科技和管理等方面的关系，考虑可利用的资源和潜在的管理决策，在资源、环境与管理约束条件下，找出最优解，使农场总利润最大化。③ 利用该类模型的相关研究多采用比较静态结构，不考虑从一个状态到另一个状态之间变化的动态过程。该类模型能够在生产计划（enterprise）选择的过程中将生产活动间的交互关系考虑在内，同时能反映资源品质的异质性以及资源约束条件，最终得到管理实践的最优集。广泛运用于澳大利亚广域农业区研究的 MIDAS（model of an integrated dryland agricultural system）就是一个典型例子。④

发展中国家家庭农业模型的目标函数一般为改善食品供给，同时需要考虑一系列小农经济的特点。⑤ 该类模型主要考虑了小规模家庭经营农业生产的资源禀赋，同时还比较了家庭消费自产农产品还是消费购自市场农产品之间的经济含义。具体的模型如 IMPACT⑥ 和 IAT⑦ 等。

---

① Kingwell, R. Managing complexity in modern farming [J]. Australian Journal of Agricultural and Resource Economics, 2011, 55: 12 – 34.

②③ Robertson, J., Pannell, J., Chalak, M. Whole-farm models: a review of recent approaches [J]. Australian Farm Business Management Journal, 2012, 9 (2): 13 – 26.

④ Kragt, M. E., Pannell, D. J., Robertson, M. J., et al. Assessing costs of soil carbon sequestration by crop-livestock farmers in Western Australia [J]. Agricultural Systems, 2012, 112: 27 – 37. Thamo, T., Kingwell, R. S., Pannell, D. J. Measurement of greenhouse gas emissions from agriculture: economic implications for policy and agricultural producers [J]. Australian Journal of Agricultural and Resource Economics, 2013, 57: 234 – 252. Kingwell, R. Managing complexity in modern farming [J]. Australian Journal of Agricultural and Resource Economics, 2011, 55: 12 – 34.

⑤ 例如家庭食品需求、非农收入、对公地的利用等。

⑥ Herrero, M. E., Gonzalez-Estrada, E., Thornton, P. K., et al. IMPACT: generic household-level databases and diagnostic tools for integrated crop-livestock analysis [J]. Agricultural Systems, 2007, 92: 240 – 265.

⑦ Lisson, S., MacLeod, N., McDonald, C., et al. A participatory, farming systems research approach to improving Bali cattle production in the smallholder crop-livestock systems of Eastern Indonesia. I. Description of process and simulation models [J]. Agricultural Systems, 2010, 103: 486 – 497.

生物物理学仿真模型通常认为资源来自外生供给，往往只有生物物理学方面的约束条件。[1] 生物物理学仿真模型一般建立在真实农场调研结果的基础之上，所以可以考虑更多农场管理实践上的细节。由于该类模型没有构建农业生产活动所面临的资源约束条件，所以不能用于对农场生产管理活动的优化。[2] 具体的应用可参见格玛雷斯等（Guimarães et al.）和莫尔等（Moore et al.）的研究。[3]

整合静态优化与动态模拟的模型将静态优化与动态仿真结合起来，以期达到取长补短的目的。在此类模型中，静态优化被用于确定在资源约束条件下农场的生产布局。此优化后的生产布局之后被用作设置实际边界。接下来，生物物理学模型可结合设置的实际边界来分析一系列如投入水平、载畜率等因素的变动所带来的结果。[4] 详细过程可参见基科沃等（Chikowo et al.）和维特布瑞德等（Whitbread et al.）的研究。[5]

结合农场实地调查的全农场模型将全农场模型与针对一些区域问题而进行的农场调研相结合来进行相关研究。[6] 此类模型可减轻只使用单一模型所带来的局限对于研究结果的负面影响。具体应用可参见科拉森斯等（Claessens et al.）的研究。[7]

自20世纪80年代以来，许多研究都利用了MIDAS来探讨澳大利亚旱地

---

[1] Department of Climate Change and Energy Efficiency. National Inventory Report 2010 Vol. 1 [R]. Department of Climate Change and Energy Efficiency, 2012.

[2] Guimarães, P. H. S., Madalena, F. E., Cezar, I. M. Comparative economics of Holstein/Gir F1 dairy female production and conventional beef cattle suckler herds – A simulation study [J]. Agricultural Systems, 2006, 88: 111 – 124.

[3] Guimarães, P. H. S., Madalena, F. E., Cezar, I. M. Comparative economics of Holstein/Gir F1 dairy female production and conventional beef cattle suckler herds-A simulation study [J]. Agricultural Systems, 2006, 88: 111 – 124. Moore, A. D., Robertson, M. J., Routley, R. Evaluation of the water use efficiency of alternative farm practices at a range of spatial and temporal scales: A conceptual framework and a modelling approach [J]. Agricultural Systems, 2011, 104: 162 – 174.

[4][6] Robertson, J., Pannell, J., Chalak, M. Whole-farm models: a review of recent approaches [J]. Australian Farm Business Management Journal, 2012, 9 (2): 13 – 26.

[5] Chikowo, R., Corbeels, M., Tittonell, P., et al. Aggregating field-scale knowledge into farm-scale models of African smallholder systems: Summary functions to simulate crop production using APSIM [J]. Agricultural Systems, 2008, 97: 151 – 166. Whitbread, A. M., Robertson, M. J., Carberry, P. S., et al. Applying farming systems simulation to the development of more sustainable smallholder farming systems in Southern Africa [J]. European Journal of Agronomy, 2010, 32: 51 – 58.

[7] Claessens, L., Stoorvogel, J. J., Antle, J. M. Ex ante assessment of dual-purpose sweet potato in the crop-livestock system of western Kenya: A minimum-data approach [J]. Agricultural Systems, 2009, 99: 13 – 22.

广域农业所面临的复杂的管理问题。① MIDAS 是一个综合考虑了广域农业在生物、管理、财务与科技方面特征的全农场生物经济静态优化模型。它为了解发生在广域农业系统内变化所造成的经济后果提供了一个有效的途径。经过相关学者的不断努力，MIDAS 的适用区域已经从最初的西澳东部小麦带扩大到西澳中部小麦带、南部海岸带、大南区（Great Southern Region）和中西区（West Midlands），以及维多利亚州西南部和新南威尔士州西部坡地区（The western slopes）②。

MIDAS 是个确定性的模型：价格与生产中的不确定性是外生的。但是，在该模型中可以改变价格与产量的水平，并分析这些改变对于农场生产计划与利润的影响。③ MIDAS 的优化目标是将农场的利润最大化。除此之外，MIDAS 也考虑了生产活动中的其他一些管理目标和行为。例如，MIDAS 允许使用者对农场经营者不同的休闲偏好进行相关设定，考虑到农民会在一月中旬开始为期数周的度假，作物收割活动可以设置在一月初结束。

MIDAS 涵盖了几百种农业活动，包括依据土壤等级改变轮作作物，依据不同牲畜等级改变饲料种类与供应，依据播种的延迟减少产量预估、现金流纪录、农机具使用等。MIDAS 的约束条件包络可利用土地面积、劳动供给、可利用资金等。MIDAS 的运行结果包括每一土地管理单元（land management unit，LMU）上轮作的选择、具体生产计划、牲畜载畜率、畜群结构、化肥施用率、预计年利润等。④

MIDAS 模型具有众多的优点，主要是能够将农业系统的生物条件和经济调节进行整合从而进行系统的分析，以及能够被运用于对一系列农业生产系

---

① Kragt, M. E., Pannell, D. J., Robertson, M. J., et al. Assessing costs of soil carbon sequestration by crop-livestock farmers in Western Australia [J]. Agricultural Systems, 2012, 112: 27–37. Thamo, T., Kingwell, R. S., Pannell, D. J. Measurement of greenhouse gas emissions from agriculture: economic implications for policy and agricultural producers [J]. Australian Journal of Agricultural and Resource Economics, 2013, 57: 234–252. Kingwell, R. Managing complexity in modern farming [J]. Australian Journal of Agricultural and Resource Economics, 2011, 55: 12–34. Pannell, D. J. Sensitivity analysis of normative economic models: theoretical framework and practical strategies [J]. Agricultural Economics, 1997, 16: 139–152.

②④ Addai, D. The economics of adaptation to climate change by broadacre farmers in Western Australia [D]. Perth: The University of Western Australia, 2013.

③ Pannell, D. J. Sensitivity analysis of normative economic models: theoretical framework and practical strategies [J]. Agricultural Economics, 1997, 16: 139–152.

统议题的研究。①

然而，MIDAS 也存在诸多局限，主要有如下方面。

其一，MIDAS 是基于一个静态优化的模型，无法考虑多年间系统的动态变化。它只能考虑两个状态点间的差异，而不能反映从一种状态变化到另一种状态的动态过程。

其二，MIDAS 没有考虑价格与生产的风险。

其三，MIDAS 操作过于复杂，不利于推广。

---

① Kingwell, R. Managing complexity in modern farming [J]. Australian Journal of Agricultural and Resource Economics, 2011, 55: 12 - 34.

参考文献

[1] 陈诗一：《工业二氧化碳的影子价格：参数化和非参数化方法》，载《世界经济》2010年第8期。

[2] 陈帅、徐晋涛、张海鹏：《气候变化对中国粮食生产的影响——基于县级面板数据的实证分析》，载《中国农村经济》2016年第5期。

[3] 陈帅：《气候变化对中国小麦生产力的影响——基于黄淮海平原的实证分析》，载《中国农村经济》2015年第7期。

[4] 陈帅：《气候变化与中国农业：粮食生产、经济影响及未来预测》，中国社会科学出版社2020年版。

[5] 陈伟娜、闫慧敏、黄河清：《气候变化压力下锡林郭勒草原牧民生计与可持续能力》，载《资源科学》2013年第5期。

[6] 程杰：《黄土高原草地植被分布与气候响应特征》，西北农林科技大学2011年博士学位论文。

[7] 程楠楠、何洪鸣、逯亚杰等：《黄土高原近52年降水时空动态特征》，载《山东农业大学学报（自然科学版）》2016年第3期。

[8] 初征、郭建平、赵俊芳：《东北地区未来气候变化对农业气候资源的影响》，载《地理学报》2017年第7期。

[9] 戴彤、王靖、赫迪等：《基于APSIM模型的气候变化对西南春玉米产量影响研究》，载《资源科学》2016年第1期。

[10] 丁勇、萨茹拉、刘朋涛等：《近40年内蒙古区域温度和降雨量变化的时空格局》，载《干旱区资源与环境》2014年第4期。

[11] 董红敏、李玉娥、陶秀萍等：《中国农业源温室气体排放与减排技术对策》，载《农业工程学报》2008年第10期。

[12] 杜文献：《气候变化对农业影响的研究进展——基于李嘉图模型的视角》，载《经济问题探索》2011第1期。

[13] 冯晓龙、刘明月、霍学喜等：《农户气候变化适应性决策对农业产出的影响效应——以陕西苹果种植户为例》，载《中国农村经济》2017 年第 3 期。

[14] 付莲莲、朱红根、周曙东：《江西省气候变化的特征及其对水稻产量的贡献——基于"气候—经济"模型》，载《长江流域资源与环境》2016 年第 4 期。

[15] 傅京燕：《环境规制与产业国际竞争力》，经济科学出版社 2006 年版。

[16] 葛亚宁、刘洛、徐新良等：《近 50a 气候变化背景下我国玉米生产潜力时空演变特征》，载《自然资源学报》2015 年第 5 期。

[17] 韩兵、苏屹、李彤等：《基于两阶段 DEA 的高技术企业技术创新绩效研究》，载《科研管理》2018 年第 3 期。

[18] 郝睿：《经济效率与地区平等：中国省际经济增长与差距的实证分析（1978 – 2003）》，载《世界经济文汇》2006 年第 2 期。

[19] 何斌、刘志娟、杨晓光等：《气候变化背景下中国主要作物农业气象灾害时空分布特征（Ⅱ）：西北主要粮食作物干旱》，载《中国农业气象》2017 年第 1 期。

[20] 何亮：《黄土高原冬小麦物候、产量和水分利用对气候变化和波动的响应》，中国科学院大学 2015 年博士学位论文。

[21] 侯玲玲、王金霞、黄季焜：《不同收入水平的农民对极端干旱事件的感知及其对适应措施采用的影响——基于全国 9 省农户大规模调查的实证分析》，载《农业技术经济》2016 年第 11 期。

[22] 侯向阳、韩颖：《内蒙古典型地区牧户气候变化感知与适应的实证研究》，载《地理研究》2011 年第 10 期。

[23] 胡慧芝、刘晓琼、王建力：《气候变化下汉中盆地水稻产量变化研究》，载《自然资源学报》2018 年第 4 期。

[24] 黄承芳、李宁、刘丽等：《气候变化下农业领域的国际文献特征与热点演变：基于 CiteSpace V 的文献计量分析》，载《中国农业气象》2019 年第 8 期。

[25] 黄小燕、李耀辉、冯建英等：《中国西北地区降水量及极端干旱气候变化特征》，载《生态学报》2015 年第 5 期。

[26] 黄泽颖、逄学思、周晓雨等：《半干旱地区马铃薯种植户适应气候

变化行为研究——基于甘肃省 362 个微观调查数据》，载《中国农业大学学报》2019 年第 1 期。

[27] IPCC：《气候变化 2007：综合报告。政府间气候变化专门委员会第四次评估报告第一、第二和第三工作组的报告》，[核心撰写组、Pachauri，R. K 和 Reisinger，A.（编辑）]，IPCC，瑞士日内瓦，2007。

[28] 贾敬敦、魏珣、金书秦：《澳大利亚发展碳汇农业对中国的启示》，载《中国农业科技导报》2012 年第 2 期。

[29] 靳乐山、魏同洋、胡振通：《牧户对气候变化的感知与适应——以内蒙古四子王旗查干补力格苏木为例》，载《自然资源学报》2014 年第 2 期。

[30] 李承政、顾海英：《气候冲击对中国县级经济的影响研究》，格致出版社 2019 年版。

[31] 李承政、李旭辉、顾海英：《气候变化计量经济学方法研究进展》，载《城市与环境研究》2019 年第 1 期。

[32] 李秋月、潘学标：《气候变化对我国北方农牧交错带空间位移的影响》，载《干旱区资源与环境》2012 年第 10 期。

[33] 李胜利、金鑫、范学珊等：《反刍动物生产与碳减排措施》，载《动物营养学报》2010 年第 1 期。

[34] 李晓燕、王彬彬：《低碳农业：应对气候变化下的农业发展之路》，载《农村经济》2010 年第 3 期。

[35] 李玉新、靳乐山：《基于社区评价的牧区适应气候变化政策改进研究——以内蒙古自治区为例》，载《农业经济问题》2013 年第 10 期。

[36] 李玥：《黄土丘陵区退耕与农业生态经济社会系统协同发展研究——以安塞县为例》，西北农林科技大学 2019 年博士学位论文。

[37] 联合国粮食与农业组织：《2016 粮食及农业状况：气候变化、农业与粮食安全》，联合国粮食与农业组织，2016 年。

[38] 林光华、陆盈盈：《气候变化对农业全要素生产率的影响及对策——以冬小麦为例》，载《农村经济》2019 年第 6 期。

[39] 刘华民、王立新、杨劼等：《气候变化对农牧民生计影响及适应性研究——以鄂尔多斯市乌审旗为例》，载《资源科学》2012 年第 2 期。

[40] 刘杰、许小峰、罗慧：《极端天气气候事件影响我国农业经济产出的实证研究》，载《中国科学：地球科学》2012 年第 7 期。

[41] 刘玉洁、陈巧敏、葛全胜等：《气候变化背景下 1981～2010 中国

小麦物候变化时空分异》，载《中国科学：地球科学》2018年第7期。

［42］马雅丽、郭建平、赵俊芳：《晋北农牧交错带作物气候生产潜力分布特征及其对气候变化的响应》，载《生态学杂志》2019年第3期。

［43］马玉平、孙琳丽、俄有浩等：《预测未来40年气候变化对我国玉米产量的影响》，载《应用生态学报》2015年第1期。

［44］毛桂莲、虎德钰、许兴等：《干旱风沙区苜蓿后茬不同轮作方式对水分利用效率和产量的影响》，载《水土保持学报》2015年第3期。

［45］闵九康：《全球气候变化和低碳农业研究》，气象出版社2011年版。

［46］宁夏回族自治区统计局：《宁夏统计年鉴2015》，中国统计出版社2015年版。

［47］农业部：《农业部关于印发〈到2020年化肥使用量零增长行动方案〉和〈到2020年农药使用量零增长行动方案〉的通知》，2015。

［48］彭佳颖：《市场激励型环境规制对企业竞争力的影响研究》，湖南大学2019年博士学位论文。

［49］蒲金涌、姚小英、王位泰：《气候变化对甘肃省冬小麦气候适宜性的影响》，载《地理研究》2011年第1期。

［50］秦大河：《中国极端天气气候事件和灾害风险管理与适应国家评估报告》，经济科学出版社2015年版。

［51］秦鹏程、姚凤梅、曹秀霞等：《利用作物模型研究气候变化对农业影响的发展过程》，载《中国农业气象》2011年第2期。

［52］秦天：《环境分权、环境规制与农业面源污染》，西南大学2020年博士学位论文。

［53］任婧宇、彭守璋、曹扬等：《1901～2014年黄土高原区域气候变化时空分布特征》，载《自然资源学报》2018年第4期。

［54］商沙沙、廉丽姝、马婷等：《近54a中国西北地区气温和降水的时空变化特征》，载《干旱区研究》2018年第1期。

［55］生态环境部：《中国应对气候变化的政策与行动2019年度报告》，2019。

［56］生态环境部：《中华人民共和国气候变化第三次国家信息通报》，2018。

［57］石晓丽、史文娇：《北方农牧交错带界线的变迁及其驱动力研究进展》，载《农业工程学报》2018年第20期。

[58] 史文娇、陶福禄、张朝：《基于统计模型识别气候变化对农业产量贡献的研究进展》，载《地理学报》2012年第9期。

[59] 史文娇、陶福禄：《非洲农业产量对气候变化响应与适应研究进展》，载《中国农业科学》2014年第16期。

[60] 孙建飞、郑聚锋、程琨等：《面向自愿减排碳交易的生物质炭基肥固碳减排计量方法研究》，载《中国农业科学》2018年第23期。

[61] 孙杨、张雪芹、郑度：《气候变暖对西北干旱区农业气候资源的影响》，载《自然资源学报》2010年第7期。

[62] 孙悦：《欧盟碳排放权交易体系及其价格机制研究》，吉林大学2018年博士学位论文。

[63] 塔西甫拉提·特依拜、丁建丽：《土地利用/土地覆盖变化研究进展综述》，载《新疆大学学报（自然科学版）》2006年第1期。

[64] 谭灵芝、王国友：《气候变化对干旱区家庭生计脆弱性影响的空间分析——以新疆于田绿洲为例》，载《中国人口科学》2012年第2期。

[65] 谭淑豪、谭文列婧、励汀郁等：《气候变化压力下牧民的社会脆弱性分析——基于内蒙古锡林郭勒盟4个牧业旗的调查》，载《中国农村经济》2016年第7期。

[66] 汤绪、杨续超、田展等：《气候变化对中国农业气候资源的影响》，载《资源科学》2011年第10期。

[67] 唐凯：《基于生物经济学的澳大利亚农业温室气体减排潜能分析》，人民出版社2018年版。

[68] 田贵良、贾琨颢、孙兴波等：《干旱事件影响下虚拟水期权契约的提出及其定价研究》，载《农业技术经济》2016年第9期。

[69] 田均良：《黄土高原生态建设环境效应研究》，气象出版社2010年版。

[70] 田磊：《变化环境下黄土高原水文气候要素数值模拟及未来预测》，西北农林科技大学2019年博士学位论文。

[71] 田云、张俊飚：《中国低碳农业发展的动态演进及收敛性研究》，载《干旱区资源与环境》2017年第3期。

[72] 汪阳洁、仇焕广、陈晓红：《气候变化对农业影响的经济学方法研究进展》，载《中国农村经济》2015年第9期。

[73] 汪阳洁、姜志德、王继军：《基于农业生态系统耦合的退耕还林工

程影响评估》，载《系统工程理论与实践》2015年第12期。

［74］王彪：《中国地方政府财政支出效率研究》，华中科技大学2012年博士学位论文。

［75］王芳：《西部循环型农业发展的理论分析与实证研究》，华中农业大学2006年博士学位论文。

［76］王继军、姜志德、连坡等：《70年来陕西省纸坊沟流域农业生态经济系统耦合态势》，载《生态学报》2009年第9期。

［77］王俊、薄晶晶、付鑫：《填闲种植及其在黄土高原旱作农业区的可行性分析》，载《生态学报》2018年第4期。

［78］王莉：《雨养农业对气候变化脆弱性的实证研究——以甘肃省华池县为例》，载《农村经济》2013年第3期。

［79］王天穷、顾海英：《我国农村能源政策以及收入水平对农户生活能源需求的影响研究》，载《自然资源学报》2017年第8期。

［80］王小彬、武雪萍、赵全胜等：《中国农业土地利用管理对土壤固碳减排潜力的影响》，载《中国农业科学》2011年第11期。

［81］王一超、郝海广、张惠远等：《农牧交错区农户生计分化及其对耕地利用的影响——以宁夏盐池县为例》，载《自然资源学报》2018年第2期。

［82］王赵琛：《24所部属高校科技成果转化效率的DEA分析》，载《科研管理》2020年第4期。

［83］魏方庆：《基于非径向距离函数DEA模型的效率评价方法研究》，中国科学技术大学2018年博士学位论文。

［84］谢伏瞻、刘雅鸣：《应对气候变化报告（2019）：防范气候风险》，社会科学文献出版社2019年版。

［85］谢姆斯叶·艾尼瓦尔、塔西甫拉提·特依拜、买买提·沙吾提等：《近50年来塔里木盆地南、北缘干湿状况变化趋势分析》，载《干旱区资源与环境》2013年第3期。

［86］胥刚：《黄土高原农业结构变迁与农业系统战略构想》，兰州大学2015年博士学位论文。

［87］徐超、杨晓光、李勇等：《气候变化背景下中国农业气候资源变化Ⅲ．西北干旱区农业气候资源时空变化特征》，载《应用生态学报》2011年第3期。

［88］薛豫南：《基于循环经济的畜禽污染治理动力机制》，大连海事大

学 2020 年博士学位论文。

[89] 延安市统计局：《延安统计年鉴 2016》，延安市统计局，2017。

[90] 闫丽、包慧娟：《奈曼旗气候变化的突变特征及其灾害效应》，载《干旱区资源与环境》2010 年第 11 期。

[91] 杨国梁：《DEA 模型与规模收益研究综述》，载《中国管理科学》2015 年第 S1 期。

[92] 杨景成、韩兴国、黄建辉等：《土壤有机质对农田管理措施的动态响应》，载《生态学报》2003 年第 4 期。

[93] 杨丽萍、王林和、秦艳等：《多伦县近 60 年气候变化分析》，载《干旱区资源与环境》2014 年第 7 期。

[94] 杨轩：《气候变化对黄土高原作物生产系统产量、水分利用及土壤养分的影响》，兰州大学 2019 年博士学位论文。

[95] 姚凤梅、秦鹏程、张佳华等：《基于模型模拟气候变化对农业影响评估的不确定性及处理方法》，载《科学通报》2011 年第 8 期。

[96] 姚玉璧、王瑞君、王润元等：《黄土高原半湿润区玉米生长发育及产量形成对气候变化的响应》，载《资源科学》2013 年第 11 期。

[97] 姚玉璧、杨金虎、肖国举等：《气候变暖对西北雨养农业及农业生态影响研究进展》，载《生态学杂志》2018 年第 7 期。

[98] 尹朝静、李谷成、高雪：《气候变化对中国粮食产量的影响——基于省级面板数据的实证》，载《干旱区资源与环境》2016 年第 6 期。

[99] 于斌斌：《产业结构调整如何提高地区能源效率？——基于幅度与质量双维度的实证考察》，载《财经研究》2017 年第 1 期。

[100] 袁鹏、程施：《我国工业污染物的影子价格估计》，载《统计研究》2011 年第 9 期。

[101] 张存厚、王明玖、李兴华等：《近 30 年来内蒙古地区气候干湿状况时空分布特征》，载《干旱区资源与环境》2011 年第 8 期。

[102] 张凡、李长生：《气候变化影响的黄土高原农业土壤有机碳与碳排放》，载《第四纪研究》2010 年第 3 期。

[103] 张四海、曹志平、张国：《保护性耕作对农田土壤有机碳库的影响》，载《生态环境学报》2012 年第 2 期。

[104] 张仲胜、李敏、宋晓林等：《气候变化对土壤有机碳库分子结构特征与稳定性影响研究进展》，载《土壤学报》2018 年第 2 期。

[105] 赵俊芳、郭建平、张艳红等：《气候变化对农业影响研究综述》，载《中国农业气象》2010年第2期。

[106] 赵凌玉、潘志华、安萍莉等：《北方农牧交错带作物耗水特征及其与气温和降水的关系——以内蒙古呼和浩特市武川县为例》，载《资源科学》2012年第3期。

[107] 赵一飞、邹欣庆、张勃等：《黄土高原甘肃区降水变化与气候指数关系》，载《地理科学》2015年第10期。

[108] 郑景云、尹云鹤、李炳元：《中国气候区划新方案》，载《地理学报》2010年第1期。

[109] 中国气象局气候变化中心：《中国气候变化蓝皮书（2019）》，2019。

[110] 中国气象局气候变化中心：《中国气候变化蓝皮书（2020）》，2020。

[111] 周洁红、唐利群、李凯：《应对气候变化的农业生产转型研究进展》，载《中国农村观察》2015年第3期。

[112] 周莉、李保国、周广胜：《土壤有机碳的主导影响因子及其研究进展》，载《地球科学进展》2005年第1期。

[113] 周鹏、周迅、周德群：《二氧化碳减排成本研究述评》，载《管理评论》2014年第11期。

[114] 周曙东、周文魁、朱红根等：《气候变化对农业的影响及应对措施》，载《南京农业大学学报：社会科学版》2010年第1期。

[115] 周志波：《环境税规制农业面源污染研究》，西南大学2019年博士学位论文。

[116] 自然资源部海洋预警司：《2019年中国海平面公报》，2020。

[117] ABARES. Agricultural Commodity Statistics 2014 [R]. ABARES, Canberra, Australia, 2014.

[118] Abberton, M. T., Marshall, A. H., Humphreys, M. W., et al. Genetic Improvement of Forage Species to Reduce the Environmental Impact of Temperate Livestock Grazing Systems [J]. Advances in Agronomy, 2008, 98: 311-355.

[119] Abid, M., Schilling, J., Scheffran, J., et al. Climate change vulnerability, adaptation and risk perceptions at farm level in Punjab, Pakistan [J]. Science of the Total Environment, 2016, 547: 447-460.

[120] ABS. Value of Agricultural Commodities Produced, Australia, 2009-

2010 [R]. Canberra: ABS, 2011.

[121] Acevedo, S., Mrkaic, M., Novta, N., Pugacheva, E., Topalova, P. The Effects of Weather Shocks on Economic Activity: What are the Channels of Impact? [J]. Journal of Macroeconomics, 2020, 103207.

[122] Addai, D. The economics of adaption to climate change by broadacre farmers in Western Australia [D]. Ph. D. thesis, School of Agricultural and Resource Economics, University of Western Australia, 2013.

[123] Adenuga, A. H., Davis, J., Hutchinson, G., Patton, M., Donnellan, T. Modelling environmental technical efficiency and phosphorus pollution abatement cost in dairy farms [J]. Science of the Total Environment, 2020, 714, 136690.

[124] Adger, W. N., Huq, S., Brown, K., et al. Adaptation to climate change in the developing world [J]. Progress in Development Studies, 2014, 3 (3): 179 - 195.

[125] Adler, N., Volta, N. Accounting for externalities and disposability: A directional economic environmental distance function [J]. European Journal of Operational Research, 2016, 250 (1): 314 - 327.

[126] Agee, M. D., Atkinson, S. E., Crocker, T. D. Child maturation, time-invariant, and time-varying inputs: their interaction in the production of child human capital [J]. Journal of Productivity Analysis, 2012, 38 (1): 29 - 44.

[127] Akinnagbe, O. M., Irohibe, I. J. Agricultural adaptation strategies to climate change impacts in Africa: a review [J]. Bangladesh Journal of Agricultural Research, 39 (3): 407 - 418.

[128] Albers, H., Gornott, C., Hüttel, S. How do inputs and weather drive wheat yield volatility? The example of Germany [J]. Food Policy, 2017, 70: 50 - 61.

[129] Aldy, J. E., Stavins, R. N. The promise and problems of pricing carbon: theory and experience [J]. Journal of Environment & Development. 2012, 21 (2): 152 - 180.

[130] Amundson, R., Guo, Y., Gong, P. Soil diversity and land use in the United States [J]. Ecosystems, 2003, 6 (5): 470 - 482.

[131] Andersen, P., Petersen, N. C. A procedure for ranking efficient

units in data envelopment analysis [J]. Management Science, 1993, 39 (10): 1261-1264.

[132] Antle, J. M., Capalbo, S. M., Mooney, S., et al. Economic analysis of agricultural soil carbon sequestration: an integrated assessment approach [J]. Journal of Agricultural and Resource Economics, 2001, 26 (2): 344-367.

[133] Antle, J. M., Capalbo, S. M., Mooney, S., et al. Sensitivity of carbon sequestration costs to soil carbon rates [J]. Environmental Pollution, 2002, 116 (3): 413-422.

[134] Antle, J. M., Capalbo, S. M., Mooney, S., et al. Spatial heterogeneity, contract design, and the efficiency of carbon sequestration policies for agriculture [J]. Journal of Environmental Economics and Management, 2003, 46 (2): 231-250.

[135] Antle, J. M., Capalbo, S. M., Paustian, K., et al. Estimating the economic potential for agricultural soil carbon sequestration in the Central United States using an aggregate econometric-process simulation model [J]. Climatic Change, 2007, 80 (1-2): 145-171.

[136] Antle, J. M., Cho, S., Tabatabaie, S. M. H., et al. Economic and environmental performance of dryland wheat-based farming systems in a 1.5℃ world [J]. Mitigation and Adaptation Strategies for Global Change, 2019, 24 (2): 165-180.

[137] Antle, J. M., Zhang, H., Mu, J. E., et al. Methods to assess between-system adaptations to climate change: dryland wheat systems in the Pacific Northwest United States [J]. Agriculture, Ecosystems & Environment, 2018, 253: 195-207.

[138] Arslan, A., Belotti, F., Lipper, L. Smallholder productivity and weather shocks: Adoption and impact of widely promoted agricultural practices in Tanzania [J]. Food Policy, 2017, 69: 68-81.

[139] Australian Department of Climate Change and Energy Efficiency. Australian national greenhouse accounts: state and territory greenhouse gas inventories [R]. Australian Department of Climate Change and Energy Efficiency, 2012.

[140] Australian Government. Emissions Reduction Fund Green Paper [R].

Commonwealth of Australia, 2014.

[141] Australian Government. Setting Australia's post-2020 target for reducing greenhouse gas emissions. Final report of the UNFCCC Taskforce [R]. Commonwealth of Australia, Canberra, 2015.

[142] Baek, C., Lee, J. D. The relevance of DEA benchmarking information and the Least-Distance Measure [J]. Mathematical & Computer Modelling, 2009, 49: 265-275.

[143] Bailes, K. L., Piltz, J. W., McNeill, D. M. In vivo digestibility of a range of silages in cattle compared with sheep [J]. Animal Production Science, 2020, 60 (5): 635-642.

[144] Bakam, I., Balana, B. B., Matthews, R. Cost-effectiveness analysis of policy instruments for greenhouse gas emission mitigation in the agricultural sector [J]. Journal of Environmental Management, 2012, 112: 33-44.

[145] Banda, D. J., Hamukwala, P., Haggblade, S., Chapoto, A. Dynamic pathways into and out of poverty: A case of smallholder farmers in Zambia (Working Paper No. 56) [Z]. Zambia: Food Security Research Project, 2011.

[146] Banker, R. D., Charnes, A., Cooper, W. W. Some models for estimating technical and scale inefficiencies in data envelopment analysis [J]. Management Science, 1984, 30 (9): 1078-1092.

[147] Banker, R. D., Conrad, R. F., Strauss, R. P. A comparative application of data envelopment analysis and translog methods: an illustrative study of hospital production [J]. Management Science, 32 (1): 30-44.

[148] Barton, L., Kiese, R., Gatter, D., et al. Nitrous oxide emissions from a cropped soil in a semi-arid climate [J]. Global Change Biology, 2008, 14: 177-192.

[149] Baumber, A., Metternicht, G., Cross, R., Ruoso, L. E., Cowie, A. L., Waters, C. Promoting co-benefits of carbon farming in Oceania: Applying and adapting approaches and metrics from existing market-based schemes [J]. Ecosystem Services, 2019, 39, 100982.

[150] Bellarby, J., Tirado, R., Leip, A., Weiss, F., Lesschen, J. P., Smith, P. Livestock greenhouse gas emissions and mitigation potential in Europe [J]. Global Change Biology, 2013, 19 (1): 3-18.

[151] Bellenger, M. J. , Herlihy, A. T. An economic approach to environmental indices [J]. Ecological Economics, 2009, 68: 2216 - 2223.

[152] Bernoux, M. , Cerri, C. C. , Cerri, C. E. P, et al. Cropping systems, carbon sequestration and erosion in Brazil, a review [J]. Agronomy for Sustainable Development, 2006, 26: 1 - 8.

[153] Bernoux, M. , Cerri, C. C. , Volkoff, B. , et al. Gaz à effet de serre et stockage du carbone par les sols, inventaire au niveau du Brésil [J]. Cahiers Agriculrures, 2005, 14: 96 - 100.

[154] Birthal, P. S. , Negi, D. S. , Khan, M. T. , Agarwal, S. Is Indian agriculture becoming resilient to droughts? Evidence from rice production systems [J]. Food Policy, 2015, 56: 1 - 12.

[155] Blancard, S. , Boussemart, J. P. , Briec, W. , Kerstens, K. Short-and long-run credit constraints in French agriculture: A directional distance function framework using expenditure-constrained profit functions [J]. American Journal of Agricultural Economics, 2006, 88 (2): 351 - 364.

[156] Bonan, G. B. , Doney, S. C. Climate, ecosystems, and planetary futures: The challenge to predict life in Earth system models [J]. Science, 2018, 359 (6375): eaam8328.

[157] Bonesmo, H. , Skjelvåg, A. O. , Henry Janzen, H. , et al. Greenhouse gas emission intensities and economic efficiency in crop production: A systems analysis of 95 farms [J]. Agricultural Systems, 2012, 110: 142 - 151.

[158] Bosch, D. J. , Stephenson, K. , Groover, G. , Hutchins, B. Farm returns to carbon credit creation with intensive rotational grazing [J]. Journal of Soil and Water Conservation, 2008, 63 (2): 91 - 98.

[159] Brümmer, B. , Glauben, T. , Lu, W. Policy reform and productivity change in Chinese agriculture: A distance function approach [J]. Journal of Development Economics, 2006, 81 (1): 61 - 79.

[160] Campbell, C. A. , Janzen, H. H. , Paustian, K. , et al. Carbon storage in soils of the North American Great Plains: effect of cropping frequency [J]. Agronomy Journal, 2005, 97 (2): 349 - 363.

[161] Capalbo, S. M. , Antle, J. M. , Mooney, S. , et al. Sensitivity of carbon sequestration costs to economic and biological uncertainties [J]. Environ-

mental Management, 2004, 33 (1): S238 - S251.

[162] Carter, M. R., Little, P. D., Mogues, T., Negatu. W. Shock, sensitivity and resilience: tracking the economic impacts of environmental disaster on assets in Ethiopia and Honduras [R]. Wisconsin, USA, 2004.

[163] Chalise, S., Naranpanawa, A. Climate change adaptation in agriculture: A computable general equilibrium analysis of land-use change in Nepal [J]. Land Use Policy, 2016, 59: 241 - 250.

[164] Challinor, A. J., Watson, J., Lobell, D. B., Howden, S. M., Smith, D. R., Chhetri, N. A meta-analysis of crop yield under climate change and adaptation [J]. Nature Climate Change, 2014, 4 (4): 287 - 291.

[165] Chambers R. G., Chung, Y., Färe, R. Benefit and distance functions [J]. Journal of Economic Theory, 1996, 70 (2): 407 - 419.

[166] Chambers, R. G., Chung, Y., Färe, R. Profit, directional distance functions, and Nerlovian efficiency [J]. Journal of Optimization Theory and Applications, 1998, 98 (2): 351 - 364.

[167] Charnes, A., Cooper, W. W., Golany, B., Seiford, L., Stutz, J. Foundations of data envelopment analysis for Pareto-Koopmans efficient empirical production functions [J]. Journal of Econometrics, 1985, 30: 91 - 107.

[168] Charnes, A., Cooper, W. W., Rhodes, E. Measuring the efficiency of decisionmaking units [J]. European journal of operational research, 1978, 2 (6): 429 - 444.

[169] Chen, H., Zhao, Y., Feng, H., Li, H., Sun, B. Assessment of climate change impacts on soil organic carbon and crop yield based on long-term fertilization applications in Loess Plateau, China [J]. Plant Soil, 2015, 390 (1 - 2): 401 - 417.

[170] Chen, Z. M., Ohshita, S., Lenzen, M., et al. Consumption-based greenhouse gas emissions accounting with capital stock change highlights dynamics of fast-developing countries [J]. Nature Communications, 2018, 9 (1): 1 - 9.

[171] Cholo, T. C., Fleskens, L., Sietz, D., Peerlings, J. Land fragmentation, climate change adaptation, and food security in the Gamo Highlands of Ethiopia [J]. Agricultural Economics, 2019, 50 (1): 39 - 49.

[172] Christensen, J. H., Hewitson, B., Busuioc, A., et al. Regional climate projections [R]. Contribution of Working Group I to the Fourth Assessment Report of the Intergovernmental Panel on Climate Change, ed. Solomon, S., Qin, D., Manning, M., et al. Cambridge, UK: Cambridge University Press, 2007.

[173] CIE. Greenhouse Gas Emissions from Agricultural Sector: Projections to 2030, CIE report to the Department of Climate Change and Energy Efficiency [R]. CIE, 2010.

[174] Clean Energy Regulator. Carbon pricing mechanism [R]. Australian Clean Energy Regulator, 2012.

[175] Clean Energy Regulator. Planning for an auction [R]. Clean Energy Regulator, Canberra, 2016.

[176] Climate Kelpie. Western Australia-weather and climate drivers [R]. Climate Kelpie, 2016.

[177] Coase, R. H. The problem of social cost [J]. The Journal of Law and Economics, 1960, 3: 1–44.

[178] Coggins, J. S., Swinton, J. R. The price of pollution: a dual approach to valuing SO_2 allowances [J]. Journal of Environmental Economics and Management, 30 (1): 58–72.

[179] Cooper, M. H., Boston, J., Bright, J. Policy challenges for livestock emissions abatement: lessons from New Zealand [J]. Climate Policy, 2013, 13 (1): 110–133.

[180] Cooper, P. J. M., Dimes, J., Rao, K. P. C., et al. Coping better with current climatic variability in the rain-fed farming systems of sub-Saharan Africa: An essential first step in adapting to future climate change? [J]. Agriculture Ecosystems & Environment, 2008, 126 (1–2): 24–35.

[181] Cooper, W. W., Park, K. S., Yu, G. An illustrative application of IDEA (imprecise data envelopment analysis) to a Korean mobile telecommunication company [J]. Operations Research, 2001, 49 (6): 807–820.

[182] DAFF. Carbon farming initiative brochure [R]. Australian Department of Agricultural, Fisheries and Forestry, 2012.

[183] Das, T. L., Saharawat, Y. S., Bhattacharyya, R, et al. Conserva-

tion agriculture effects on crop and water productivity, profitability and soil organic carbon accumulation under a maize-wheat cropping system in the North-western Indo-Gangetic Plains [J]. Field Crops Research, 2018, 215: 222 - 231.

[184] De Cara, S., Houzé, M., Jayet, P. A. Methane and nitrous oxide emissions from agriculture in the EU: A spatial assessment of sources and abatement costs [J]. Environmental and Resource Economics, 2005, 32 (4): 551 - 583.

[185] De Cara, S., Jayet, P. A. Marginal abatement costs of greenhouse gas emissions from European agriculture, cost effectiveness, and the EU non-ETS burden sharing agreement [J]. Ecological Economics, 2011, 70 (9): 1680 - 1690.

[186] Dervaux, B., Leleu, H., Minvielle, E., Valdmanis, V., Aegerter, P., Guidet, B. Performance of French intensive care units: A directional distance function approach at the patient level [J]. International Journal of Production Economics, 2009, 120 (2): 585 - 594.

[187] Deschenes, O., Greenstone, M. The economic impacts of climate change: evidence from agricultural output and random fluctuations in weather [J]. American Economic Review, 2007, 97 (1): 354 - 385.

[188] Desjardins, R. L., Kulshreshtha, S. N., Junkins, B., et al. Canadian greenhouse gas mitigation options in agriculture [J]. Nutrient Cycling in Agroecosystems, 2001, 60 (1 - 3): 317 - 326.

[189] Deutsch, C. A., Tewksbury, J. J., Tigchelaar, M., et al. Increase in crop losses to insect pests in a warming climate [J]. Science, 2018, 361 (6405): 916 - 919.

[190] Dong, H., Li, Y., Tao, X., Li, N., Zhu, Z. China's greenhouse gas emissions from agricultural activities and its mitigation strategy [J]. Transactions of the CSAE, 2008, 24 (10): 269 - 273.

[191] Doole, G. J., Bathgate, A. D., Robertson, M. J. Labour scarcity restricts the potential scale of grazed perennial plants in the Western Australian Wheatbelt [J]. Animal Production Science, 2009, 49: 883 - 893.

[192] Doran-Browne, N., Wootton, M., Taylor, C., Eckard, R. Offsets required to reduce the carbon balance of sheep and beef farms through carbon sequestration in trees and soils [J]. Animal Production Science, 2018, 58 (9):

1648 – 1655.

[193] Dumbrell, N. P., Kragt, M. E., Gibson, F. L. What carbon farming activities are farmers likely to adopt? A best-worst scaling survey [J]. Land Use Policy, 2016, 54: 29 – 37.

[194] Eldoma, I. M., Li, M., Zhang, F., Li, F. M. Alternate or equal ridge-furrow pattern: Which is better for maize production in the rain-fed semi-arid Loess Plateau of China? [J]. Field Crops Research, 2016, 191: 131 – 138.

[195] Emrouznejad, A., Yang, G. L. A survey and analysis of the first 40 years of scholarly literature in DEA: 1978 – 2016 [J]. Socio-Economic Planning Sciences, 2018, 61: 4 – 8.

[196] European Commission. 20 20 by 2020: Europe's Climate Change Opportunity COM (2008) 30 final Commission of the European Communities [R]. Brussels, Belgium, 2008.

[197] Fahad, S., Wang, J. Farmers' risk perception, vulnerability, and adaptation to climate change in rural Pakistan [J]. Land Use Policy, 2018, 79: 301 – 309.

[198] Fan, Y., Wu, J., Xia, Y., Liu, J. Y. How will a nationwide carbon market affect regional economies and efficiency of CO_2 emission reduction in China? [J]. China Economic Review, 2016, 38: 151 – 166.

[199] Fang, J., Yu, G., Liu, L., Hu, S., Chapin III, F. S. Climate change, human impacts, and carbon sequestration in China [J]. Proceedings of the National Academy of Sciences, 2018, 115 (16): 4015 – 4020.

[200] FAO, IFAD, UNICEF, WFP, WHO. The state of food security and nutrition in the world 2018: building climate resilience for food security and nutrition [R]. FAO, Rome, 2018.

[201] Färe, R., Grosskopf, S. Directional distance functions and slacks-based measures of efficiency: Some clarifications [J]. European Journal of Operational Research, 2010, 206 (3): 702 – 702.

[202] Färe, R., Grosskopf, S., Noh, D. W., Weber, W. Characteristics of a polluting technology: Theory and practice [J]. Journal of Econometrics, 2005, 126 (2): 469 – 492.

[203] Färe, R., Grosskopf, S., Pasurka, C. A. Environmental production functions and environmental directional distance functions [J]. Energy, 2007, 32 (7): 1055 – 1066.

[204] Färe, R., Grosskopf, S., Weber, W. L. Shadow prices and pollution costs in US agriculture [J]. Ecological economics, 2006, 56 (1): 89 – 103.

[205] Farquharson, R., Abadi, A., Finlayson, J., et al. EverFarm®– Climate adapted perennial-based farming systems for dryland agriculture in southern Australia [R]. National Climate Change Adaptation Research Facility, Gold Coast Australia, 2013.

[206] Feng, X., Fu, B., Piao, S., et al. Revegetation in China's Loess Plateau is approaching sustainable water resource limits [J]. Nature Climate Change, 2016, 6 (11): 1019 – 1022.

[207] Fiala, N. Meeting the demand: an estimation of potential future greenhouse gas emissions from meat production [J]. Ecological Economics, 2008, 67 (3): 412 – 419.

[208] Fisher, A. C., Hanemann, W. M., Roberts, M. J., et al. The economic impacts of climate change: evidence from agricultural output and random fluctuations in weather: comment [J]. American Economic Review, 2012, 102 (7): 3749 – 3760.

[209] Flannery, T, Beale, R., Hueston, G, The critical Decade: International Action on Climate Change [R]. Commonwealth of Australia Department of Climate Change and Energy Efficiency, 2012.

[210] Flugge, F., Abadi, A. Farming carbon: an economic analysis of agroforestry for carbon sequestration and dryland salinity reduction in Western Australia [J]. Agroforestry Systems, 2006, 68: 181 – 192.

[211] Food and Agriculture Organization of the United Nations. FAO Strategy on Climate Change [R]. Food and Agriculture Organization of the United Nations, Rome, 2017.

[212] Frank, S., Havlík, P., Stehfest, E., et al. Agricultural non-CO_2 emission reduction potential in the context of the 1.5℃ target [J]. Nature Climate Change, 2019, 9 (1): 66 – 72.

[213] Franzluebbers, A. J. Achieving soil organic carbon sequestration with conservation agricultural systems in the southeastern United States [J]. Soil Science Society of AmericaJournal, 2010, 74 (2): 347-357.

[214] Frei, F. X., Harler, P. T. Projections onto efficient frontiers: Theoretical and computational extensions to DEA [J]. Journal of Productivity Analysis, 1999, 11 (3): 275-300.

[215] Freibauer, A. Regionalised inventory of biogenic greenhouse gas emissions from European agriculture [J]. European Journal of Agronomy, 2003, 19: 135-160.

[216] Fu, B., Liu, Y., Lü, Y., et al. Assessing the soil erosion control service of ecosystems change in the Loess Plateau of China [J]. Ecological Complexity, 2011, 8 (4): 284-293.

[217] Fu, B., Wang, Y., Lu, Y., He, C., Chen, L., Song, C. The effects of land-use combinations on soil erosion: a case study in the Loess Plateau of China [J]. Progress in Physical Geography, 2009, 33 (6): 793-804.

[218] Fu, W., Huang, M., Gallichand, J., Shao, M. Optimization of plant coverage in relation to water balance in the Loess Plateau of China [J]. Geoderma, 2012, 173-174: 134-144.

[219] Fynn, A. J., Alvarez, P., Brown, J. R., et al. Soil carbon sequestration in US rangelands: issues paper for protocol development [R]. Environmental Defense Fund, 2009.

[220] Gan, Y., Liang, C., Hamel, C., Cutforth, H., Wang, H. Strategies for reducing the carbon footprint of field crops for semiarid areas. A review [J]. Agronomy for Sustainable Development, 2011, 31 (4): 643-656.

[221] Gan, Y. T., Campbell, C. A., Jansen, H. H., et al. Carbon input to soil by oilseed and pulse crops in semiarid environment [J]. Agriculture, Ecosystem & Environment, 2009, 132: 290-297.

[222] Garnaut, R. The Garnaut Climate Change Review [M]. Cambridge: Cambridge University Press, 2008.

[223] Ghahramani, A., Bowran, D. Transformative and systemic climate change adaptations in mixed crop-livestock farming systems [J]. Agricultural Systems, 2018, 164: 236-251.

[224] Ghahramani, A., Kingwell, R. S., Maraseni, T. N. Land use change in Australian mixed crop-livestock systems as a transformative climate change adaptation [J]. Agricultural Systems, 2020, 102791.

[225] Gong, C., Tang, K., Zhu, K., Hailu, A. An optimal time-of-use pricing for urban gas: A study with a multi-agent evolutionary game-theoretic perspective [J]. Applied Energy, 2016, 163: 283 – 294.

[226] González-Estrada, E., Rodriguez, L. C., Walen, V. K., et al. Carbon sequestration and farm income in West Africa: Identifying best management practices for smallholder agricultural systems in northern Ghana [J]. Ecological Economics, 2008, 67 (3): 492 – 502.

[227] Government of India. Assessment of climate change over the Indian Region. A report of the Ministry of Earth Sciences (MoES), Government of India [R]. Springer Nature Singapore, Singapore, 2020.

[228] Grace, P. R., Antle, J., Aggarwal, P. K., et al. Soil carbon sequestration and associated economic costs for farming systems of the Indo-Gangetic Plain: A meta-analysis. Agriculture, Ecosystems & Environment [J], 2012, 146 (1): 137 – 146.

[229] Grace, P. R., Antle, J., Ogle, S., et al. Soil carbon sequestration rates and associated economic costs for farming systems of south-eastern Australia [J]. Australian Journal of Soil Research, 2010, 48: 1 – 10.

[230] Grains Research and Development Corporation. 2015 Farm Gross Margin Guide: A gross margin template for crop and livestock enterprises [R]. Grains Research and Development Corporation, Canberra, 2015.

[231] Grosskopf, S., Hayes, K. J., Taylor, L. L., Weber, W. L. Anticipating the consequences of school reform: A new use of DEA [J]. Management Science, 1999, 45 (4): 608 – 620.

[232] Guan, K., Sultan, B., Biasutti, M., Baron, C., Lobell, D. B. Assessing climate adaptation options and uncertainties for cereal systems in West Africa [J]. Agricultural and Forest Meteorology, 2017, 232: 291 – 305.

[233] Hailu, A. APEAR: A package for productivity and efficiency analysis in R (version0.1) [Z]. UWA, 2013. Available from URL: http://ahailu.are.uwa.edu.au

[234] Hailu, A., Chambers, R. G. A Luenberger soil-quality indicator [J]. Journal of Productivity Analysis, 2012, 38 (2): 145-154.

[235] Hailu, A., Durkin, J., Sadler, R., Nordblom, T. L. Agent-based modelling study of shadow, saline water table management in the Katanning catchment, Western Australia [R]. Research Report for RIRDC Project No. PRJ-000578, 2011.

[236] Hailu, A., Veeman, T. S. Non-parametric productivity analysis with undesirable outputs: an application to the Canadian pulp and paper industry [J]. American Journal of Agricultural Economics, 2001, 83 (3): 605-616.

[237] Halkos, G. E., Tzeremes, N. G. A conditional directional distance function approach for measuring regional environmental efficiency: Evidence from UK regions [J]. European Journal of Operational Research, 2013, 227 (1): 182-189.

[238] Hampf, B., Krüger, J. J. Technical efficiency of automobiles-A non-parametric approach incorporating carbon dioxide emissions [J]. Transportation Research Part D: Transport & Environment, 2014, 33: 47-62.

[239] Harris, D., Orr, A. Is rainfed agriculture really a pathway from poverty? [J] Agricultural Systems, 2014, 123: 84-96.

[240] Havlík, P., Valin, H., Mosnier, A., et al. Crop productivity and the global livestock sector: Implications for land use change and greenhouse gas emissions [J]. American Journal of Agricultural Economics, 2012, 95 (2): 442-448.

[241] Hawkins, J., Ma, C., Schilizzi, S., Zhang, F. China's changing diet and its impacts on greenhouse gas emissions: an index decomposition analysis [J]. Australian Journal of Agricultural and Resource Economics, 2018, 62 (1): 45-64.

[242] Herrero, M., Henderson, B., Havlík, P., et al. Greenhouse gas mitigation potentials in the livestock sector [J]. Nature Climate Change, 2016, 6 (5): 452-461.

[243] Herrero, M., Thornton, P. K., Notenbaert, A. M., et al. Smart investments in sustainable food production: revisiting mixed crop-livestock systems [J]. Science, 2010, 327 (5967): 822-825.

[244] Herridge, D. F., Peoples, M. B., Boddey, R. M. Global inputs of biological nitrogen fixation in agricultural systems [J]. Plant and Soil, 2008, 311 (1-2): 1-18.

[245] Hoang, M. H., Do, T. H., Pham, M. T., van Noordwijk, M., Minang, P. A. Benefit distribution across scales to reduce emissions from deforestation and forest degradation (REDD+) in Vietnam [J]. Land Use Policy, 2013, 31: 48-60.

[246] Hoffmann, I. Livestock genetic diversity and climate change adaptation [A]. In: Rowlinson, P., Steele, M., Nefzaoui, A. (Eds.), Proceedings of the International Conference Livestock and Global Climate Change [C]. Hammamet, Tunisia, 2008.

[247] Holper, P. N. Climate Change, Science Information Paper: Australian Rainfall: Past, Present and Future [R]. CSIRO, 2011.

[248] Huang, J., Yu, H., Guan, X., Wang, G., Guo, R. Accelerated dryland expansion under climate change [J]. Nature Climate Change, 2016, 6 (2): 166-171.

[249] Huffman, W. E., Jin, Y., Xu, Z. The economic impacts of technology and climate change: new evidence from US corn yields [J]. Agricultural Economics, 2018, 49 (4): 463-479.

[250] Hughes, L., Steffen, W. The Critical Decade: Climate Change Science, Risks and Responses [R]. Climate Commission Secretariat, Australia, 2013.

[251] Hunt, C. Economy and ecology of emerging markets and credits for bio-sequestered carbon on private land in tropical Australia [J]. Ecological Economics, 2008, 66 (2): 309-318.

[252] IPCC. 2006 Guidelines for National Greenhouse Gas Inventories. Intergovernmental Panel on Climate Change [R]. National Greenhouse Gas Inventories Programme, IPCC, 2006.

[253] IPCC. Global Warming of 1.5℃ [R]. IPCC, 2018.

[254] IPCC. IPCC Fifth Assessment Report (AR5) [M]. Cambridge: Cambridge University Press, 2014.

[255] IPCC. Managing the Risks of Extreme Events and Disasters to Advance Climate Change Adaptation. A Special Report of Working Groups Ⅰ and Ⅱ of the

Intergovernmental Panel on Climate Change [R]. Cambridge University Press, Cambridge, UK, 2012.

[256] IPCC. Summary for policymakers [M]. Cambridge, United Kingdom and New York, USA: Cambridge University Press, 2007.

[257] Islam, N., Xayavong, V., Kingwell, R. Broadacre farm productivity and profitability in south-western Australia [J]. Australian Journal of Agricultural and Resource Economics, 2014, 58 (2): 147-170.

[258] Jackson, J., Choudrie, S., Thistlethwaite, G., et al. UK Greenhouse Gas Inventory, 1990 to 2007: Annual Report for submission under the Framework Convention on Climate Change [R]. UK Department of Energy and Climate Change, 2009.

[259] Ji, Y., Ranjan, R., Burton, M. A bivariate probit analysis of factors affecting partial, complete and continued adoption of soil carbon sequestration technology in rural China [J]. Journal of Environmental Economics and Policy, 2017, 6 (2): 153-167.

[260] John, M. The economics of dryland salinity management in a low-rainfall environment of Western Australia [D]. Ph. D. thesis, School of Agricultural and Resource Economics, University of Western Australia, 2004.

[261] Kaiser, H. M., Riha, S. J., Wilks, D. S., et al. A farm-level analysis of economic and agronomic impacts of gradual climate warming [J]. American Journal of Agricultural Economics, 1993, 75 (2): 387-398.

[262] Kalkuhl, M., Wenz, L. The impact of climate conditions on economic production. Evidence from a global panel of regions [J]. Journal of Environmental Economics and Management, 2020, 103, 102360.

[263] Kao, C. Network data envelopment analysis: A review [J]. European Journal of Operational Research, 2014, 239 (1): 1-16.

[264] Khakbazan, M., Mohr, R. M., Derksen, D. A., et al. Effects of alternative management practices on the economics, energy and GHG emissions of a wheat-pea cropping system in the Canadian prairies [J]. Soil and Tillage Research, 2009, 104 (1): 30-38.

[265] Khanal, U., Wilson, C., Hoang, V. N., et al. Farmers' adaptation to climate change, its determinants and impacts on rice yield in Nepal [J].

Ecological Economics, 2018, 144: 139 - 147. Agriculture, Ecosystems & Environment, 2016, 220: 164 - 174.

[266] Khataza, R. R., Hailu, A., Kragt, M. E., Doole, G. J. Estimating shadow price for symbiotic nitrogen and technical efficiency for legume-based conservation agriculture in Malawi [J]. Australian Journal of Agricultural and Resource Economics, 2017, 61 (3): 462 - 480.

[267] Kingwell, R., Jeanne, R. M., Hailu, A. A longitudinal analysis of some Australian broadacre farms' greenhouse gas emissions, farming systems and efficiency of production [J]. Agricultural Systems, 2016, 146: 120 - 128.

[268] Kingwell, R. S. Using mathematical programming to model farm management under price and seasonal uncertainty: An analysis of stabilisation policies for wheat and wool [D]. Ph. D. thesis, University of Western Australia, 1996.

[269] Kirkegaard, J., Christen, O., Krupinsky, J., Layzell, D. Break crop benefits in temperate wheat production [J]. Field Crops Research, 2008, 107 (3): 185 - 195.

[270] Kragt, M. E., Pannell, D. J., Robertson, M. J., Thamo, T. Assessing costs of soil carbon sequestration by crop-livestock farmers in Western Australia [J]. Agricultural Systems, 2012, 112: 27 - 37.

[271] Kroon, F. J., Kuhnert, P. M., Henderson, B. L., et al. River loads of suspended solids, nitrogen, phosphorus and herbicides delivered to the Great Barrier Reef lagoon [J]. Marine Pollution Bulletin, 2012, 65 (4 - 9): 167 - 181.

[272] Kumar, S. Environmentally sensitive productivity growth: A global analysis using Malmquist-Luenberger index [J]. Ecological Economics, 2006, 56 (2): 280 - 293.

[273] Kumar, S., Managi, S., Jain, R. K. CO_2 mitigation policy for Indian thermal power sector: Potential gains from emission trading [J]. Energy Economics, 2020, 86, 104653.

[274] Lal, R. Soil carbon sequestration impacts on global climate change and food security [J]. Science, 2004, 304: 1623.

[275] Lal, R. Soil management and restoration for C sequestration to mitigate the accelerated greenhouse effect [J]. Progress in Environmental Science,

1999, 1 (4): 307 - 326.

[276] Lamb J. D., Tee, K. H. Data envelopment analysis models of investment funds [J]. European Journal of Operational Research, 2012, 216 (3): 687 - 696.

[277] Lee, C. Y. Meta-data envelopment analysis: Finding a direction towards marginal profit maximization [J]. European Journal of Operational Research, 2014, 237 (1): 207 - 216.

[278] Leleu, H. Shadow pricing of undesirable outputs in nonparametric analysis [J]. European Journal of Operational Research, 2013, 231 (2): 474 - 480.

[279] Lewis, K. A., Tzilivakis, J., Green, A., et al. The climate change mitigation potential of an EU farm: towards a farm-based integrated assessment [R]. University of Hertfordshire, UK, 2010.

[280] Li, X., Philp, J., Cremades, R., Roberts, A., Liang, H., Li, L., Yu, Q. Agricultural vulnerability over the Chinese Loess Plateau in response to climate change: exposure, sensitivity, and adaptive capacity [J]. Ambio, 2016, 45 (3): 350 - 360.

[281] Li, Y., Barton, L., Chen, D. Simulating response of N2O emissions to fertiliser N application and climatic variability from a rain-fed and wheat-cropped soil in Western Australia [J]. Journal of the Science of Food and Agriculture, 2011, 92: 1130 - 1143.

[282] Li, Z., Liu, W., Zhang, X., Zheng, F. Assessing the site-specific impacts of climate change on hydrology, soil erosion and crop yields in the Loess Plateau of China [J]. Climatic Change, 2011, 105 (1 - 2): 223 - 242.

[283] Liu, D. L., Chan, K. Y., Conyers, M. K. Simulation of soil organic carbon under different tillage and stubble management practices using the Rothamsted carbon model [J]. Soil & Tillage Research, 2009, 104: 65 - 73.

[284] Liu, H., Owens, K. A., Yang, K., Zhang, C. Pollution abatement costs and technical changes under different environmental regulations [J]. China Economic Review, 2020, 62, 101497.

[285] Liu, J., Liu, M., Zhuang, D., Zhang, Z., Deng, X. Study on spatial pattern of land-use change in China during 1995 - 2000 [J]. Science in

China Series D: Earth Sciences, 2003, 46 (4): 373 - 384.

[286] Liu, J. Y., Feng, C. Marginal abatement costs of carbon dioxide emissions and its influencing factors: A global perspective [J]. Journal of Cleaner Production, 2018, 170: 1433 - 1450.

[287] Liu, L. C., Wu, G. The effects of carbon dioxide, methane and nitrous oxide emission taxes: An empirical study in China [J]. Journal of Cleaner Production, 2017, 142: 1044 - 1054.

[288] Liu, W., Sang, T. Potential productivity of the Miscanthus energy crop in the Loess Plateau of China under climate change [J]. Environmental Research Letters, 2013, 8 (4), 044003.

[289] Liu, Y. Introduction to land use and rural sustainability in China [J]. Land Use Policy, 2018, 74: 1 - 4.

[290] Liu, Y., Fang, F., Li, Y. Key issues of land use in China and implications for policy making [J]. Land Use Policy, 2014, 40: 6 - 12.

[291] Lobell, D. B., Bänziger, M., Magorokosho, C., Vivek, B. Nonlinear heat effects on African maize as evidenced by historical yield trials [J]. Nature Climate Change, 2011, 1 (1): 42 - 45.

[292] Lobell, D. B., Cahill, K. N., Field, C. B. Historical effects of temperature and precipitation on California crop yields [J]. Climatic Change, 2007, 81 (2): 187 - 203.

[293] Lobell, D. B., Hammer, G. L., McLean, G., et al. The critical role of extreme heat for maize production in the United States [J]. Nature Climate Change, 2013, 3 (5): 497 - 501.

[294] Lobell, D. B., Roberts, M. J., Schlenker, W., et al. Greater sensitivity to drought accompanies maize yield increase in the US Midwest [J]. Science, 2014, 344 (6183): 516 - 519.

[295] Luenberger, D. G. Benefit functions and duality [J]. Journal of Mathematical Economics, 2004, 21 (5): 461 - 481.

[296] Machefert, S. E., Dise, N. B., Goulding, KWT et al. Nitrous oxide emission from a range of land uses across Europe [J]. Hydrology and Earth System Sciences, 2002, 6: 325 - 337.

[297] Maestre, F. T., Quero, J. L. Gotelli, N. J. Plant species richness

and ecosystem multifunctionality in global drylands [J]. Science, 2012, 335 (6065): 214-218.

[298] Marsh, S. P., Pannell, D. J., Lindner, R. K. Does agricultural extension pay? A case study for a new crop, lupins, in Western Australia [J]. Agricultural Economics, 2004, 30 (1): 17-30.

[299] Martínez, C. C., Angás, P., Lampurlanés, J. Long-term yield and water use efficiency under various tillage systems in Mediterranean rainfed conditions [J]. Annals of Applied Biology, 2007, 150 (3): 293-305.

[300] Mayberry, D., Bartlett, H., Moss, J., Davison, T., Herrero, M. Pathways to carbon-neutrality for the Australian red meat sector [J]. Agricultural Systems, 2019, 175: 13-21.

[301] Maziotis, A., Villegas, A., Molinos-Senante, M. The cost of reducing unplanned water supply interruptions: A parametric shadow price approach [J]. Science of the Total Environment, 2020, 719, 137487.

[302] McKinsey & Company. Australian Cost Curve for GHG Reduction [R]. McKinsey & Company, Australia, 2008.

[303] Meat and Livestock Australia. Detailed saleyard report sheep and lamb: Katanning (03 Feb 2016) [R]. Meat and Livestock Australia, Sydney, 2016.

[304] Meier, E. A., Thorburn, P. J., Kragt, M. E., et al. Greenhouse gas abatement on southern Australian grains farms: Biophysical potential and financial impacts [J]. Agricultural Systems, 2017, 155: 147-157.

[305] Mendelsohn, R., Nordhaus, W. D., Shaw, D. The impact of global warming on agriculture: a Ricardian analysis [J]. The American economic review, 1994: 753-771.

[306] Messing, I., Chen, L., Hessel, R. Soil conditions in a small catchment on the Loess Plateau in China [J]. Catena, 2003, 54 (1-2): 45-58.

[307] Metcalf, G. E. A proposal for a US carbon tax swap: An equitable tax reform to address global climate change [R]. Discussion Paper 2007-12, Hamilton Project, Brookings Institute, 2007.

[308] Molinos-Senante, M., Sala-Garrido, R. How much should customers

be compensated for interruptions in the drinking water supply? [J]. Science of The Total Environment, 2017, 586: 642 – 649.

[309] Monteny, G. J., Bannink, A., Chadwick, D. Greenhouse gas abatement strategies for animal husbandry [J]. Agriculture, Ecosystems & Environment, 2006, 112 (2): 163 – 170.

[310] Mooney, S., Antle, J., Capalbo, S., et al. Influence of project scale and carbon variability on the costs of measuring soil carbon credits [J]. Environmental Management, 2004, 33 (1): S252 – S263.

[311] Moorby, J. M., Chadwick, D. R., Scholefield, D., et al. A review of best practice for reducing greenhouse gases [R]. Defra project report AC0206, 2007.

[312] Moore, G. (ed.). Soil guide: A handbook for understanding and managing agricultural soils [R]. Department of Agriculture, Western Australia. Bulletin No. 4343, 2001.

[313] Morrison, D. A., Kingwell, R. S., Pannell, D. J., Ewing, M. A. A mathematical programming model of a crop-livestock farm system [J]. Agricultural Systems, 1986, 20 (4): 243 – 268.

[314] Mosavi, S. H., Alipour, A., Shahvari, N. Liberalizing energy price and abatement cost of emissions: Evidence from Iranian agro-environment [J]. Journal of Agricultural Science and Technology, 2017, 19 (3): 511 – 523.

[315] Ndjeunga, J., Savadogo, K. Changes in rural household livelihood strategies and outcomes in Burkina Faso [A]. In: Freeman, H. A., Rohrbach, D. D., Ackello-Ogutu, C. (Eds.), Targeting Agricultural Research for Development in the Semi-Arid Tropics of Sub-Saharan Africa. Proceedings of a Workshop held at International Center for Research in Agroforestry [C]. Nairobi, Kenya, 2002.

[316] Nguyen, H. Q. Analyzing the economies of crop diversification in rural Vietnam using an input distance function [J]. Agricultural Systems, 2017, 153: 148 – 156.

[317] Njuki, E., Bravo-Ureta, B. E. The economic costs of environmental regulation in U. S. dairy farming: A directional distance function approach [J]. American Journal of Agricultural Economics, 2015, 97 (4): 1087 – 1106.

[318] Nolan, S., Unkovich, M., Yuying, S., Lingling, L., Bellotti, W. Farming systems of the Loess Plateau, Gansu Province, China [J]. Agriculture, Ecosystems & Environment, 2008, 124 (1): 13 - 23.

[319] O'Hara, P., Freney, J., Ulyatt, M. Abatement of agricultural non-carbon dioxide greenhouse gas emissions [R]. Report for the Ministry of Agriculture and Forestry, 2003.

[320] Oenema, O., Witzke, H. P., Klimont, Z., et al. Integrated assessment of promising measures to decrease nitrogen losses from agriculture in EU-27 [J]. Agriculture, Ecosystems & Environment, 2009, 133 (3): 280 - 288.

[321] Oum, T. H., Pathomsiri, S., Yoshida, Y. Limitations of DEA-based approach and alternative methods in the measurement and comparison of social efficiency across firms in different transport modes: An empirical study in Japan [J]. Transportation Research Part E: Logistics and Transportation Review, 2013, 57: 16 - 26.

[322] Packett, R. Rainfall contributes 30% of the dissolved inorganic nitrogen exported from a southern Great Barrier Reef river basin [J]. Marine Pollution Bulletin, 2017, 121 (1 - 2): 16 - 31.

[323] Packett, R., Dougall, C., Rohde, K., Noble, R. Agricultural lands are hot-spots for annual runoff polluting the southern Great Barrier Reef lagoon [J]. Marine Pollution Bulletin, 2009, 58 (7): 976 - 986.

[324] Parks, P. J., Hardie, I. W. Least-cost forest carbon reserves: Cost-effective subsidies to convert marginal agricultural land to forests [J]. Land economics, 1995, 71 (1): 122 - 136.

[325] Parliament of the Commonwealth of Australia. Carbon Farming Initiative Amendment Bill, Explanatory Memorandum [R]. Parliament of the Commonwealth of Australia, Canberra, 2014.

[326] Pendell, D. L., Williams, J. R., Boyles, S. B., et al. Soil carbon sequestration strategies with alternative tillage and nitrogen sources under risk [J]. Applied Economic Perspectives and Policy, 2007, 29 (2): 247 - 268.

[327] Picazo-Tadeo, A. J., Beltrán-Esteve, M., Gómez-Limón, J. A. Assessing eco-efficiency with directional distance functions [J]. European Journal of Operational Research, 2012, 220 (3): 798 - 809.

[328] Picazo-Tadeo, A. J., Reig-Martínez, E., Hernández-Sancho, F. Directional distance functions and environmental regulation [J]. Resource & Energy Economics, 2005, 27 (2): 131-142.

[329] Pigou, A. The Economics of Welfare [M]. London: The Macmillan Company, 1920.

[330] Piontek, F., Kalkuhl, M., Kriegler, E., Schultes, A., Leimbach, M., Edenhofer, O., Bauer, N. Economic growth effects of alternative climate change impact channels in economic modeling [J]. Environmental and Resource Economics, 2019, 73 (4): 1357-1385.

[331] Powell, J. W., Welsh, J. M., Eckard, R. J. An irrigated cotton farm emissions case study in NSW, Australia [J]. Agricultural Systems, 2017, 158: 61-67.

[332] Powlson, D. S., Stirling, C. M., Thierfelder, C., et al. Does conservation agriculture deliver climate change mitigation through soil carbon sequestration in tropical agro-ecosystems? [J]. Agriculture, Ecosystems & Environment, 2016, 220: 164-174.

[333] Rahn, E., Liebig, T., Ghazoul, J., et al. Opportunities for sustainable intensification of coffee agro-ecosystems along an altitudinal gradient on Mt. Elgon, Uganda [J]. Agriculture, Ecosystems & Environment, 2018, 263: 31-40.

[334] Rajaniemi, M., Mikkola, H., Ahokas, J. Greenhouse gas emissions from oats, barley, wheat and rye production [J]. Agronomy Research, 2011, 9 (1): 189-195.

[335] Ray, S. C., Mukherjee, K. Decomposition of cost competitiveness in US manufacturing: Some state-by-state comparisons [J]. Indian Economic Review, 2000, 35 (2): 133-153.

[336] Reeves, D. W. The role of soil organic matter in maintaining soil quality in continuous cropping systems [J]. Soil and Tillage Research, 1997, 43 (1): 131-167.

[337] Roberts, M. J., Schlenker, W., Eyer, J. Agronomic weather measures in econometric models of crop yield with implications for climate change [J]. American Journal of Agricultural Economics, 2012, 95 (2): 236-243.

[338] Robertson, J. , Pannell, J. , Chalak, M. Whole-farm models: a review of recent approaches. Australian Farm Business Management Journal, 2012, 9 (2): 13 - 26.

[339] Robertson, S. M. , Friend, M. A. Performance of sheep systems grazing perennial pastures. 4. Simulated seasonal variation and long-term production [J]. Animal Production Science, 2020, 60 (3): 423 - 435.

[340] Rockström, J. , Falkenmark, M. Agriculture: increase water harvesting in Africa [J]. Nature, 2015, 519 (7543): 283 - 285.

[341] Rojas-Downing, M. M. , Nejadhashemi, A. P. , Harrigan, T. , et al. Climate change and livestock: Impacts, adaptation, and mitigation [J]. Climate Risk Management, 2017, 16: 145 - 163.

[342] Rust, S. , Star, M. The cost effectiveness of remediating erosion gullies: a case study in the Fitzroy [J]. Australasian Journal of Environmental Management, 2018, 25 (2): 233 - 247.

[343] Salazar-Espinoza, C. , Jones, S. , Tarp, F. Weather shocks and cropland decisions in rural Mozambique [J]. Food Policy, 2015, 53: 9 - 21.

[344] Samuelson, P. A. The pure theory of public expenditure [J]. The Review of Economics and Statistics, 1954, 36 (4): 387 - 389.

[345] Sanderman, J. , Farquharson, R. , Baldock, J. Soil carbon sequestration potential: a review for Australian agriculture [R]. CSIRO Sustainable Agriculture National Research Flagship, 2010.

[346] Schirmer, J. , Parsons, M. , Charalambou, C. Gavran, M. Socio-economic impacts of plantation forestry in the Great Southern region of WA [R]. Forest and Wood Products Research and Development Corporation, Melbourne, VIC, 2005.

[347] Schlenker, W. , Hanemann, W. M. , Fisher, A. C. Will US agriculture really benefit from global warming? Accounting for irrigation in the hedonic approach [J]. The American Economic Review, 2005, 95 (1): 395 - 406.

[348] Schlenker, W. , Roberts, M. J. Nonlinear temperature effects indicate severe damages to US crop yields under climate change [J]. Proceedings of the National Academy of sciences, 2009, 106 (37): 15594 - 15598.

[349] Schlesinger, W. H. , Andrews, J. A. , Soil respiration and the global

carbon cycle [J]. Biogeochemistry, 2000, 48 (1): 7-20.

[350] Seiford, L. M., Zhu, J. Profitability and marketability of the top 55 US commercial banks [J]. Management Science, 1999, 45 (9): 1270-1288.

[351] Sexton, T. R., Silkman, R. H., Hogan, A. J. Data envelopment analysis: Critique and extensions [J]. New Directions for Evaluation, 2010, 1986 (32): 73-105.

[352] Sheng, Y., Xu, X. The productivity impact of climate change: Evidence from Australia's Millennium drought [J]. Economic Modelling, 2019, 76: 182-191.

[353] Shephard, R. W. Theory of Cost and Production Functions [M]. Princeton: Princeton University Press, 1970.

[354] Sidibé, Y., Foudi, S., Pascual, U., Termansen, M. Adaptation to climate change in rainfed agriculture in the global south: soil biodiversity as natural insurance [J]. Ecological Economics, 2018, 146: 588-596.

[355] Simar, L., Vanhems, A., Wilson, P. W. Statistical inference for DEA estimators of directional distances [J]. European Journal of Operational Research, 2012, 220 (3): 853-864.

[356] Simelton, E., Fraser, E. D., Termansen, M., et al. The socioeconomics of food crop production and climate change vulnerability: a global scale quantitative analysis of how grain crops are sensitive to drought [J]. Food Security, 2012, 4 (2): 163-179.

[357] Singh, B. Mineralogical and chemical characteristics of soils from south-western Australia [D]. Ph. D. thesis, School of Agriculture, University of Western Australia, 1991.

[358] Skidmore, S., Santos, P., Leimona, B. Targeting REDD+: An Empirical Analysis of Carbon Sequestration in Indonesia [J]. World Development, 2014, 64: 781-790.

[359] Smith, P, Martino, D, Cai, Z, et al. Greenhouse gas mitigation in agriculture [J]. Philosophical Transactions Royal Society of London Biological Sciences, 2008, 363: 789-813.

[360] Smith, P. Agricultural greenhouse gas mitigation potential globally, in Europe and in the UK: What have we learnt in the last 20 years? [J]. Global

Change Biology, 2012, 18 (1): 35 -43.

[361] Sommer, S. G., Petersen, S. O., Sørensen, P, et al. Methane and carbon dioxide emissions and nitrogen turnover during liquid manure storage [J]. Nutrient Cycling in Agroecosystems, 2007, 78: 27 -36.

[362] Srinivasarao, C., Lal, R., Kundu, S., Babu, M. P., Venkateswarlu, B., Singh, A. K. Soil carbon sequestration in rainfed production systems in the semiarid tropics of India [J]. Science of the Total Environment, 2012, 487: 587 -603.

[363] Stavins, R. N. The cost of carbon sequestration: a revealed-preference approach [J]. The American Economic Review, 1999, 89 (4): 994 -1009.

[364] Stevanovic, M., Popp, A., Bodirsky, B. L., et al. Mitigation strategies for greenhouse gas emissions from agriculture and land-use change: consequences for food prices [J]. Environmental Science & Technology, 2017, 51 (1): 365 -374.

[365] Su, Z., Zhang, J., Wu, W., et al. Effects of conservation tillage practices on winter wheat water-use efficiency and crop yield on the Loess Plateau, China [J]. Agricultural Water Management, 2007, 87: 307 -314.

[366] Tang, K. Economic analysis of the potential for greenhouse gas mitigation through carbon farming in Australia's broadacre agricultural sector [D]. Ph. D. thesis, School of Agricultural and Resource Economics, University of Western Australia, 2016.

[367] Tang, K., Gong, C., Wang, D. Reduction potential, shadow prices, and pollution costs of agricultural pollutants in China [J]. Science of the Total Environment, 2016, 541: 42 -50.

[368] Tang, K., Hailu, A. Smallholder farms' adaptation to the impacts of climate change: Evidence from China's Loess Plateau [J]. Land Use Policy, 2020, 91, 104353.

[369] Tang, K., Hailu, A., Kragt, M. E., Ma, C. Marginal abatement costs of greenhouse gas emissions: broadacre farming in the Great Southern Region of Western Australia [J]. Australian Journal of Agricultural and Resource Economics, 2016, 60 (3): 459 -475.

[370] Tang, K., Hailu, A., Kragt, M. E., Ma, C. The response of

broadacre mixed crop-livestock farmers to agricultural greenhouse gas abatement incentives [J]. Agricultural Systems, 2018, 160: 11 - 20.

[371] Tang, K., Hailu, A., Yang, Y. Agricultural chemical oxygen demand mitigation under various policies in China: A scenario analysis [J]. Journal of Cleaner Production, 2020, 250, 119513.

[372] Tang, K., He, C., Ma, C., Wang, D. Does carbon farming provide a cost-effective option to mitigate GHG emissions? Evidence from China [J]. Australian Journal of Agricultural and Resource Economics, 2019, 63 (3): 575 - 592.

[373] Tang, K., Kragt, M. E., Hailu, A., Ma, C. Carbon farming economics: What have we learned? [J]. Journal of Environmental Management, 2016, 172: 49 - 57.

[374] Thamo, T. Climate change in Western Australia agriculture: A bio-economic and policy analysis [D]. Ph. D. thesis, School of Agricultural and Resource Economics, University of Western Australia, 2017.

[375] Thamo, T., Addai, D., Pannell, D. J., et al. Climate change impacts and farm-level adaptation: Economic analysis of a mixed cropping-livestock system [J]. Agricultural Systems, 2017, 150: 99 - 108.

[376] Thamo, T., Kingwell, R. S., Pannell, D. J. Measurement of greenhouse gas emissions from agriculture: Economic implications for policy and agricultural producers [J]. Australian Journal of Agricultural and Resource Economics, 2013, 57: 234 - 252.

[377] The Climate Institute, Policy Brief-Australia and the Future of the Kyoto Protocol [R]. The Climate Institute, 2012.

[378] Thornton, P. K., Steeg, J. V. D., Notenbaert, A., et al. The impacts of climate change on livestock and livestock systems in developing countries: A review of what we know and what we need to know [J]. Agricultural Systems, 2009, 101 (3): 113 - 127.

[379] Tone, K. A slacks-based measure of efficiency in data envelopment analysis [J]. European Journal of Operational Research, 2001, 130 (3): 498 - 509.

[380] Trinh, T. Q., Rañola Jr, R. F., Camacho, L. D., Simelton, E.

Determinants of farmers' adaptation to climate change in agricultural production in the central region of Vietnam [J]. Land Use Policy, 2018, 70: 224 – 231.

[381] Tschakert, P. The costs of soil carbon sequestration: an economic analysis for small-scale farming systems in Senegal [J]. Agricultural Systems, 2004, 81 (3): 227 – 253.

[382] Tsunekawa, A., Liu, G., Yamanaka, N., Du, S. Restoration and Development of the Degraded Loess Plateau, China [M]. Springer, Japan, 2014.

[383] United Nations Framework Convention on Climate Change (UNFCCC), Kyoto Protocol (United Nations: Germany), A decision support framework for greenhouse accounting on Australian dairy, beef or grain farms [R]. UNFCCC, 2009.

[384] van der Weerden, T. J., Luo, J., de Klein, C. A. M, et al. Disaggregating nitrous oxide emission factors for ruminant urine and dung deposited onto pastoral soils [J]. Agriculture, Ecosystems & Environment, 2011, 141: 426 – 436.

[385] van Valkengoed, A. M., Steg, L. Meta-analyses of factors motivating climate change adaptation behaviour [J]. Nature Climate Change, 2019, 9 (2): 158 – 163.

[386] Villano, R., Fleming, E., Fleming, P. Evidence of farm-level synergies in mixed-farming systems in the Australian Wheat-Sheep Zone [J]. Agricultural Systems, 2010, 103 (3): 146 – 152.

[387] Wahab, M. I. M., Wu, D., Lee, C. G. A generic approach to measuring the machine flexibility of manufacturing systems [J]. European Journal of Operational Research, 2008, 186 (1): 137 – 149.

[388] Wang, K., Che, L., Ma, C., Wei, Y. The shadow price of CO_2 emissions in China's iron and steel industry [J]. Science of the Total Environment, 2017, 598: 272 – 281.

[389] Wang, K., Xian, Y., Lee, C. Y., Wei, Y. M., Huang, Z. On selecting directions for directional distance functions in a non-parametric framework: A review [J]. Annals of Operations Research, 2019, 278: 43 – 76.

[390] Wang, S., Chu, C., Chen, G., Peng, Z., Li, F. Efficiency

and reduction cost of carbon emissions in China: A non-radial directional distance function method [J]. Journal of Cleaner Production, 2016, 113: 624-634.

[391] Wang, S., Fu, B., Chen, H., Liu, Y. Regional development boundary of China's Loess Plateau: Water limit and land shortage [J]. Land Use Policy, 2018, 74: 130-136.

[392] Wang, S., Fu, B., Piao, S., et al. Reduced sediment transport in the Yellow River due to anthropogenic changes [J]. Nature Geoscience, 2016, 9 (1): 38.

[393] Wang, W., Koslowski, F., Nayak, D. R., et al. Greenhouse gas mitigation in Chinese agriculture: distinguishing technical and economic potentials [J]. Global Environmental Change, 2014, 26: 53-62.

[394] Wang, Y., Zhang, X., Huang, C. Spatial variability of soil total nitrogen and soil total phosphorus under different land uses in a small watershed on the Loess Plateau, China [J]. Geoderma, 2009, 150 (1-2): 141-149.

[395] Wei, X., Zhang, N. The shadow prices of CO_2 and SO_2 for Chinese coal-fired power plants: A partial frontier approach [J]. Energy Economics, 2020, 85, 104576.

[396] Western Australian Government. How Australia accounts for agricultural greenhouse gas emissions [R]. Western Australian Government, Perth, WA, 2017.

[397] Wiebe, K., Lotze-Campen, H., Sands, R., et al. Climate change impacts on agriculture in 2050 under a range of plausible socioeconomic and emissions scenarios [J]. Environmental Research Letters, 2015, 10 (8), 085010.

[398] Wirsenius, S., Hedenus, F., Mohlin, K. Greenhouse gas taxes on animal food products: rationale, tax scheme and climate mitigation effects [J]. Climatic Change, 2011, 108 (1-2): 159-184.

[399] World Bank. Health Nutrition and Population Statistics [R]. World Bank, Washington DC, 2016.

[400] World Bank. Looking beyond the horizon: how climate change impacts and adaptation responses will reshape agriculture in Eastern Europe and Central Asia [R]. World Bank, Washington, D. C., 2013.

[401] Wossen, T., Berger, T., Haile, M. G., Troost, C. Impacts of climate variability and food price volatility on household income and food security of farm households in East and West Africa [J]. Agricultural Systems, 2018, 163: 7-15.

[402] WRI. Climate Analysis Indicators Tool: WRI's Climate Data Explorer [Z]. World Resources Institute, Washington, D. C., 2014.

[403] Wu, D., Li, S., Liu, L., Lin, J., Zhang, S. Dynamics of pollutants' shadow price and its driving forces: An analysis on China's two major pollutants at provincial level [J]. Journal of Cleaner Production, 2020. DOI: 10.1016/j.scitotenv, 2020, 136690

[404] Wu, J., Ma, C. 2019. The convergence of China's marginal abatement cost of CO_2: An emission-weighted continuous state space approach [J]. Environmental and resource economics, 2019, 72 (4): 1099-1119.

[405] Wu, X., Zhang, J., You, L. Marginal abatement cost of agricultural carbon emissions in China: 1993-2015 [J]. China Agricultural Economic Review, 2018, 10 (4): 558-571.

[406] Xie, H., Chen, Q., Wang, W., He, Y. Analyzing the green efficiency of arable land use in China [J]. Technological Forecasting and Social Change, 2018, 133: 15-28.

[407] Xin, Z., Yu, X., Li, Q., Lu, X. X. Spatiotemporal variation in rainfall erosivity on the Chinese Loess Plateau during the period 1956-2008 [J]. Regional Environmental Change, 2011, 11 (1): 149-159.

[408] Xu, J., Li, B., Wu, D. Rough data envelopment analysis and its application to supply chain performance evaluation [J]. International Journal of Production Economics, 2009, 122 (2): 628-638.

[409] Xu, W., Yin, Y., Zhou, S. Social and economic impacts of carbon sequestration and land use change on peasant households in rural China: A case study of Liping, Guizhou Province [J]. Journal of Environmental Management, 2007, 85 (3): 736-745.

[410] Yang, F., Wu, D. D., Liang, L., O'Neill, L. Competition strategy and efficiency evaluation for decision making units with fixed-sum outputs [J]. European Journal of Operational Research, 2011, 212 (3): 560-569.

[411] Yang, L., Tang, K., Wang, Z., An, H., Fang, W. Regional eco-efficiency and pollutants' marginal abatement costs in China: A parametric approach [J]. Journal of Cleaner Production, 2017, 167: 619-629.

[412] Yuan, Z. Q., Yu, K. L., Epstein, H., et al. Effects of legume species introduction on vegetation and soil nutrient development on abandoned croplands in a semi-arid environment on the Loess Plateau, China [J]. Science of the Total Environment, 2016, 541: 692-700.

[413] Zelek, C. A., Shively, G. E. Measuring the opportunity cost of carbon sequestration in tropical agriculture [J]. Land Economics, 2003, 79 (3): 342-354.

[414] Zhang, B., He, C., Burnham, M., Zhang, L. Evaluating the coupling effects of climate aridity and vegetation restoration on soil erosion over the Loess Plateau in China [J]. Science of the Total Environment, 2016, 539: 436-449.

[415] Zhang, W. F., Dou, Z. X., He, P., et al. New technologies reduce greenhouse gas emissions from nitrogenous fertilizer in China [J]. Proceedings of the National Academy of Sciences, 2013, 110 (21): 8375-8380.

[416] Zhang, Y., Wang, R., Wang, S., et al. Effects of different subsoiling frequencies incorporated into no-tillage systems on soil properties and crop yield in dryland wheat-maize rotation system [J]. Field Crops Research, 2017, 209: 151-158.

[417] Zhang, Z., Baranzini, A. What do we know about carbon taxes? An inquiry into their impacts on competitiveness and distribution of income [J]. Energy Policy, 2004, 32 (4): 507-518.

[418] Ziervogel, G., Cartwright, A., Tas, A., et al. Climate change and adaptation in African agriculture [R]. Stockholm Environment Institute, 2008.

[419] Zofio, J. L., Pastor, J. T., Aparicio, J. The directional profit efficiency measure: On why profit inefficiency is either technical or allocative [J]. Working Papers in Economic Theory, 2010, 40 (3): 257-266.